The Gnat Is Older than Man

GLOBAL ENVIRONMENT AND HUMAN AGENDA

✣

Other Titles by Christopher D. Stone

Earth and Other Ethics

Where the Law Ends: The Social Control of Corporate Behavior

Should Trees Have Standing? Toward Legal Rights for Natural Objects

Law, Language, and Ethics

The Gnat Is Older than Man

GLOBAL ENVIRONMENT
AND HUMAN AGENDA

✣

CHRISTOPHER D. STONE

PRINCETON UNIVERSITY PRESS

PRINCETON, NEW JERSEY

Library of Congress Cataloging-in-Publication Data

Stone, Christopher D.
The gnat is older than man : Global environment and human agenda /
Christopher D. Stone
p.   cm.
Includes index.
ISBN 0-691-03250-5 (acid-free)
1. Environmental policy. 2. Environmental Law. 3. Human ecology.
4. Natural resources—Moral and ethical aspects. I. Title.
HC79.E5S75   1993
363.7—dc20   92-21139

This book has been composed in Linotron Berkeley

Printed on recycled paper

✤ *Dedication* ✤

The world was made for man, though he
was the latecomer among its creatures.
This was design. He was to find all
things ready for him. God was the host
who prepared dainty dishes, set the table,
and then led his guest to his seat. At
the same time man's late appearance on earth
is to convey an admonition of humility.
Let him beware of being proud, lest he invite
the retort that the gnat is older than he.
—*Talmud*

In a hospital in Boston, my father, seventy years a journalist and riding the crest of a successful volume on Socrates, spoke with enthusiasm of getting out of bed and into a new book he was planning but declined to discuss. He died never revealing what it was to have been about.

At the time, it seemed to me the saddest thing about Dad's death, and yet the most glorious about his life, that he should have carried to the very end the feeling that he had one more book in him. Knowing that was enough. I gave the mystery of his intentions no more thought.

In the summer of 1992 on the way to my family home in Washington, D.C., I stopped off in Princeton to discuss with Princeton University Press among other things a title for this book. I continued on to Washington with the suggestion that I might "look for something from the Bible, or something like that."

The house that I had grown up in, governed now by my mother alone, had changed. Of the huge library which once furnished and inspirited it, fewer than a thousand volumes stood guard. (On his deathbed, six days after Dad had authorized me to cut off the *Boston Globe*, he summoned my sister Celia, his eldest, and at that time when most men's minds turn to final guidance for their progeny, delivered instructions on the disposition of his books.) But more than a few things remained. For one, on an otherwise empty shelf, a batch of notes on notecards. From these it became clear that the intended book was to have gone back to Adam and Creation. ("For out of Zion shall go forth the law." What had this nonobservant deeply religious atheist intended to do with that?) And there I discovered, browsing nearby among the surviving volumes, the passage set out above, ". . . *the gnat is older than he.*" What did it all mean? But of course, that was it, the title I was after—and the opportunity for the lastborn to draw from the tangle of unresolved mysteries and love that bind creator and created a dedication of this work to my father, I. F. Stone, and the literary tradition from which he came.

✤ Contents ✤

✢ Preface ✢

A Parable for Our Time?

In March 1988, a giant oil tanker, the Exxon Valdez, *ran onto a reef in Prince William Sound, Alaska, fouling thousands of miles of sea and beach with nearly 10 million gallons of crude oil. While the ship was veering toward this fate, its captain was down in his cabin, his "judgment," on one account, "impaired by alcohol."[1] The helm had been assigned to a third mate, who was unlicensed to pilot the channel. The lookout was not on the bow. The ship was crammed with state-of-the-art navigational equipment, including a depth alarm that was not working. The ship's steering had been entrusted to its Sperry Marine SRP-2000 automatic pilot, a device normally reserved for the open seas; as the ship plowed closer to the reef, no one thought to turn it off despite three warning lights indicating it was on and an override switch that would easily have taken it out of operation. The National Traffic Safety Board was to fault the Coast Guard for failing to have tracked the ship long enough on its radar and to have warned it back onto course by radio. Indeed, it may have been, as much as anything else, a misplaced faith in the capacity of high technology to compensate for human error that lulled the* Valdez *to ruin.[2]*

And, of course, the tragedy does not end there. In disregard of good advice, much of the effort to clean the beaches of oil was entrusted to steam and hot-water washing. The effects of the hot water on the food chain are believed to have been far more destructive than had the oil simply been left in place. Much of the teeming vital "lower life" was buried, smothered, and cooked.[3]

In its way, the wreck of the Exxon Valdez *and the impairment of the Alaskan environment is a parable for our planet. Everyone on board has something else to do, other than worry about where the whole thing, our earth, is heading. Those nominally in charge have their hands full just reviewing the legion of little day-to-day chores that must be done: the terrestrial counterparts of getting the brass polished, the boilers stoked, the mess served. We hope that those who are supposed to be steering know what they are doing. Or perhaps some imminent breakthrough in technology will rescue us from the deficit of human wisdom on the bridge. But nothing calms the uneasy feeling that the whole earthly vessel is adrift, uncaptained and in peril.*

What can be done?

A Personal Passage

I BEGAN WORK on this book in the spring of 1988, in the wake, as it were, of the *Valdez* incident. The intervening five years provided more than the ordinary (that is to say, wonderful) fulfillments of research and writing. They offered a boon of false starts. Some of the proposals I had planned to make when I embarked, such as Environmental Guardianships (chapter 4) and the Global Commons Trust Fund (chapter 9), I came to hold more firmly. But many ideas that I had held originally turned out to be wrong or had to be recast. These revisions in my thinking can be gathered into a preface of instructive errors.

1. "The global environmental crisis will bring us all together."

One of the themes with which I began was that we could regard the environmental crisis as others had regarded the threat of nuclear destruction: "Pollution has . . . become a binding force among nations," one oceans expert had proclaimed.[4] Another international lawyer echoed this hope in an article in the *Israeli Law Review* in 1977—that in the face of the environmental challenge, "The overriding law of human solidarity is bound to cut through narrow ideas of national sovereignty."[5] By the 1980s the world had become inspired by the joint efforts of American and Soviet cosmonauts working together in space as part of a United Nations Environment

Programme (UNEP) project to monitor global environmental trends. The symbolism of the new perspective on the planet, not to mention the actual endeavor, was riveting.

I am still attracted to this vision of an environment-driven global harmony, and hope that it is fulfilled. But I am increasingly uncertain. The sometimes rancorous Earth Summit in Rio de Janeiro in June 1992 highlighted terrible and growing conflicts among nations that the common threats to the environment are only intensifying. At worst, rather than rouse us into global solidarity, the environmental crisis may simply provide more things to fight about. I found that I could not write a book about the environment that did not confront the rich-poor tensions, politically and philosophically.

2. *"Global problems require global solutions."*

Many of the problems we face transcend national boundaries. A molecule of carbon dioxide emitted into the atmosphere in Warsaw has the same effect on the earth's heat blanket as a molecule released in Rio. Tearing up the Antarctic, or using the oceans as a dumping ground for radioactive wastes, deprives all the world's peoples of a share of their common heritage. It is natural to suppose that what we need are global solutions: great global Earth Summits, multinational treaties, a Global Environmental Fund, perhaps even a Global Environmental Agency modeled after our own Environmental Protection Agency, but with earth-spanning powers of investigation and command.

Now, I am not so sure. Like most people who teach international law, I have a soft spot for world order. Global, or at least large-number multilateral agreements, are often advantageous. But they are rarely if ever indispensable. And sometimes the costs of collecting all humankind into great corrective frameworks is not worth it in terms of frictions, delays, bureaucratic inertia, and the price of purchasing cooperation. Many problems, even those as apparently "global" as climate change and biodiversity, lend themselves to a wider variety of bilateral and unilateral responses than one is first inclined to assume (chapter 7). A noble inclination for global cooperation should not distract us from more modestly scaled and realistically achievable local undertakings.

3. "Our problems all stem from Human Greed (or from the Judeo-Christian ethic, or population or capitalism)."

I have become less persuaded that any of the invariably cited spiritual failings of humankind—or even all of them together—do justice to the rich complexity of our predicaments. Genesis, it is charged, suffused the Judeo-Christian tradition with a manly scent that has aroused environmentalists and feminists into a common affront ("ecofeminism"), the one repulsed that she should be "his help meet," the other that the Amazon and dolphins should be his "dainty dishes." Indeed. But plenty of preclassical and non-Western civilizations made a mess of their environments without any benefit of the Bible.[6] The Western standard of living, to which just about the entire world is aspiring (I am not sure it is illuminating to call that "greed"), brings with it a lot of bad baggage and shameful levels of waste. But we should not be blind to how much degradation springs from ignorance, need, and even from devotion: how else to explain the desertification that comes from people just trying to scratch from already overtaxed soil a marginal life for their families? In all events, placing the blame on spiritual shortcomings pins too much hope on spiritual reform in an era stamped by material factors, technology, and impersonal institutions.

Not that I dismiss the spiritual and ethical dimensions; they cap this work as its final and to me most important chapter. But if the world is going to be changed, it is going to be changed by people who can get past talking up a reform of the human spirit and lessons to be learned from Buddhism, and put in the effort required to understand fisheries quotas, pollution taxes, trade barriers, and what the International Court of Justice can and cannot do.

In the same vein, while population growth puts dreadful pressures on local environments in many parts of the world, the link to environmental degradation is not straightforward. Most of the serious,

globe-spanning maladies are the by-product of industrial civilizations that display low or even negative rates of population growth. In
other words, checking population would relieve pressure on the environment (and would be warranted if only to raise hundreds of
millions of people into a condition in which they could enjoy a spir-
itually and physically acceptable life). But even if population levels
off, some of the most serious environmental problems will, with expanding industrialization, remain to be dealt with. Birth control pills
can't control toxic wastes.

4. "Prevention is the best remedy."

We are instructed from childhood that it is better to be safe than
sorry, that a stitch in time saves nine, that you cannot put the tooth-
paste back in the tube. All of these homilies make a considerable
amount of sense—in law as in life. Traditional legal remedies do not
provide any relief until after the world has changed in ways we do
not like. With the benefit of hindsight, when we think we know why
something went wrong, we guiltily imagine that some measure we
could have required in advance would have averted the whole disaster: better radar or sonar instrumentation, better hiring policies,
stronger or redundant hulls, and so forth. On the other hand, the
problem with preventive regulations is that they carry the risk of
overregulation. They are necessarily deployed before we know
which ship on which route is fated to crash from what causes; yet
each mandate is destined to apply across the board to an entire class
of ships and sea lanes, raising costs even on the many voyages that
could have been plied without incident.

Put otherwise (and not to imply any judgment on the particulars
of the *Exxon Valdez* tragedy), an ounce of prevention may be worth
a pound of cure; but by the same quaint token, to pay a pound in
prevention to get only an ounce of cure is a distinctly bad buy. And
even while I come out favoring increased reliance on preventive arrangements in circumstances where there are real cost benefits or
special reasons to favor a "premium" for risk avoidance, the question
whether to lean toward prevention or cure is not one that can be

answered generally. The proponents of adaptation—of wait-and-see and then respond-as-called-for—cannot marshall as large an army of maxims and fables (and there is something unnerving in that!). But in many circumstances where one first would not have imagined it, adaptation—combined with some insurance-style risk-spreading—may be a superior policy.

5. *"The Earth is going to hell in a climate-changing basket."*

When I started writing, I was persuaded by a barrage of news stories that we faced a global-warming crisis primed to unleash mighty typhoons, parch the world's grainlands, and swamp us under rising tides of ocean bloated by melted ice caps.

For most of mankind, this is bad news. But for an academic author, it offered some possibility of cutting my losses. Books about the end of the world (as of history, ideology, nature, or just of the booming stock market, for that matter) I am told sell. I thought I would join in the growing chorus, or commerce, emphasizing what international law might contribute to stave off doom.

While I pressed ahead to write the doleful text, I dispatched a student researcher to cull the scientific literature to trail me with the supportive footnotes. (That is how it is done.) In this spirit, having recited in a draft the popular menace that the polar ice caps were ready to melt on us and so on, I waited for the authoritative backing to materialize in memos. I waited in vain. The deeper into the better authorities we fished, the vaguer and more qualified the projections we landed. One view (now, I believe, the prevailing one) turned out to be that global warming, rather than raise sea level through melted ice caps, would more likely thicken the caps and thereby to some extent counteract any rise in sea level caused by anticipated thermal expansion of the waters.* And it wasn't just the ice-sheet data that

---

* While this sounds counterintuitive, as explained more fully in chapter 1, the seas, like anything else, would expand when heated and therefore rise; but at the same time, there would be a conflicting ocean-lowering effect, since more warmth would trigger more evaporation and precipitation, transferring a massive volume of water out of the oceans and storing it on the polar caps in the form of snow.

were being blown out of proportion. Over the space of the few years that I have been following the research developments, all of the original, highly publicized projections of climate change variables have without exception crept back to much more modest levels than in the original scare stories.

This brought to mind a nagging recollection. In the mid-1970s I had been one of a bevy of academics recruited by the U.S. government as foot soldiers in the Energy War, charged with figuring out how to deal with the well-publicized fact that the world was "running out of oil"—had, in fact, only twenty years' supply left.* Inevitably, the closer I got to people who knew what they were talking about, the more doubtful the crisis appeared. Indeed, we now know that even though world consumption has gone up in the past twenty years,** further discoveries (of which we are assured few more were forthcoming) and advances in oil field technology had pushed proved reserves worldwide from 640 billion barrels in 1977 to a record one trillion barrels by mid-1992—a forty-year inventory, virtually higher than at any time since such records began to be kept in the 1930s.

Yet in the 1970s the plan to save America from empty wells was moving ahead of the facts with such momentum that we were at the brink of forcing the economy to convert from oil to coal (in which the United States was well endowed) at immense expense in terms of economic dislocation, dirty air, and, we can say in retrospect, greenhouse complications. The déjà-vu troubled me. Global warming was not to be dismissed. But I had been burned once.

---

* The story with natural gas supplies is even more of the same: the supplies are appearing faster than we consume them. In fact, in the past twenty years the emphasis of public policy has shifted from finding inexpensive fossil fuels to weaning ourselves of them, largely for environmental reasons. Not all the efforts of us energy warriors was wasted. The Energy War jolted the U.S. into conservation measures and R&D to develop alternative energy sources. But some of the more zealous battles, like the threat to make industry yank out a trillion dollar investment in natural oil and gas-fueled boilers and replace them with coal-fired equipment, I'm relieved our generals lost.

** From the 50 million barrels a day (mbd) range in the 1970s, to 69 mbd in 1992.

This is not the place to go into detail. I try, particularly in chapter 1, to summarize the global-warming debate as evenhandedly as possible. In advance, I do not regard the dangers as trivial in the least, and throughout the book am mindful of opportunities to combat them. But it seems quite clear that people are being misled to regard the perils of climate change as overshadowing all other environmental problems. Before Rio, it was proclaimed that a climate change convention was the highest priority on the agenda. (This insistence led to diplomatic tussles that in turn undid efforts to reach agreements on other matters such as biodiversity and forests.) When U.S. resistance to accept toothy constraints on carbon emissions undermined agreement, environmentalists proclaimed the meeting a failure and, identifying a willingness to expend trillions of dollars to prevent a peril that may never eventuate with caring for the environment, denounced the United States for "not caring."[7]

I find it very distressing for so much of the environmental movement to hitch its wagon to climate change. To do so risks eventually discrediting the movement; it distracts energies away from much more significant issues; and it aggravates misunderstanding among the nations of the world.

Put aside, for a moment, the problems that I will return to of oceans, species, forests, biodiversity, the polar regions, and desertification. Worldwide, as many as 10 million people are infected with the AIDS virus, and 40 million may be by the year 2000.[8] The United Nations International Children's Emergency Fund (UNICEF) estimates that in the 1990s, 100 million people, 27,000 a day, principally children living in the Third World, will die of treatable diseases[9]—many of which result from dirty water, the world's most truly murderous environmental problem.[10] Malaria alone is said (with some likely exaggeration) to claim perhaps a million a year.[11] Measles, whooping cough, and tetanus, all inoculable, take over 3 million young lives a year;[12] diarrhea and respiratory infections claim nearly 16,000 a day.[13]

At the meeting in Rio, Mahbubul Haq, a special adviser to the United Nations Development Program, put well the feelings I have grown to share: "Although global warming has yet to kill a single human being and may not do so for centuries, it has received enormous attention and resources. At the same time, silent emergencies that are killing people every day—the fact that one point three billion people live in desertifying areas—do not attract the same kind of screaming headlines and well-funded action plans."[14]

The same sort of point might even be made in regard to the ozone shield. Action to phase out ozone-depleting agents was remarkably rapid as these things go, and is frequently offered (somewhat unrealistically, we shall see) as a model for global cooperation. But it offers something else to think about. In 1987 the Environmental Protection Agency (EPA), in pressing for controls, claimed that failure to restrict the use of chlorofluorocarbons (CFCs) would result in 3,782,000 skin cancer deaths by the year 2075 in the United States alone, a figure that, if you work it out, comes to an average of 43,000 per year.[15] Such forecasts based upon past experience have to be considered, of course, highly unreliable, both because of likely adjustments in human behavior over an eighty-year stretch—for example, increased awareness of the problem and consequent countermeasures such as sun lotions—and anticipated improvements in cancer therapy over the time horizons involved. But aside from the uncertainty, it is intriguing, considering the relatively high level of publicity the problem has drawn, to note how *low* the speculative peril is compared with the number who are indubitably dying *each day now*, with hardly an eyelash batted, from unsanitary water, malaria, measles, tetanus, and other diseases.[16]

It makes one wonder: What forces motivate the Western press and determine so much of the human agenda?

I do not have all the answers. Part of it may be as simple as "out of sight, out of mind." But one cannot dismiss a sort of insidious, conscience-numbing racism: How much are we influenced by the fact that the preponderance of the treatable unsanitary deaths occur among dark-skinned Third World children—and that the ozone-connected risk of skin cancer is highest among white populations living in high latitudes, and lowest among dark-skinned populations in midlatitudes? Such racial and North-South differences in outlook and agenda were the sources of many of the misunderstandings and frictions at the 1992 Rio Earth Summit (and are examined in chapters 5 and 10).

It is also obvious that climate change and ozone depletion have gotten people to open their pockets to environmental groups more than dirty water, and even whales, ever did. It has been, as one environmental newsletter to which I subscribe cheerily headlined, a "Boom Time for Environmentalists."[17] But the explanation and motivations are not simply mercenary. Given the choice, almost anyone would rather say he or she is working on a solution to global warming than on malaria. Malaria sounds old hat.*

And it must be said, too, that a certain ambivalence, even complicity, in the Third World's attitudes toward climate change reinforces the exaggerations. The less developed countries (LDCs) are of course concerned about global warming. Third World economies are in many ways more vulnerable to climate change than the highly diversified and industrialized economies of the First World. But even when we account for the LDCs having, typically, more urgent items on their agenda, they are aware that alarm over global warming (warranted or not) among the rich is promoting an uneasy environmentalist-Third World alliance calculated to transfer wealth to the poor under the worthy but vague banner of "sustainable development." That theme, too, is one that runs throughout this book.

At this point, someone is bound to interject that nothing can be done so long as the press can claim "The Third World Suffers under a Mountain of Debt"—and there is no debt relief in sight. But while this is true, and crucial to what follows, it, too, is only part-true. Mountains of debt or not, the Third World finds the wherewithal to purchase $40 billion in weaponry each year. The United States may be flagging in autos, but it ranked number one in arms sales—over $65.8 billion in 1990 and 1991 combined, $32.6 billion of it to

---

* When I suggested this at a presentation in Cleveland, a doctor in the audience impressed upon me that, scientifically speaking, there is no challenge left to be discovered about malaria. That turns out to be a slight exaggeration. See Louis Miller, "The Challenge of Malaria," reporting ongoing malaria research in *Science* 257 (1992): 36–38, but it is true that eliminating malaria and other such diseases is as much of a test for political will as for medical science.

Third World countries.[18] This success in arms sales was achievable largely because the government obliges our domestic arms manufacturers with subsidies (directly and with a wink and a nod at cost overruns and embargo violations) and by fostering sales to their customers who would otherwise be at risk of blowing one another up with some other nation's weaponry. Thus we have a deputy secretary of state firing off a classified memo to all U.S. embassies abroad urging that defense firms be given more help marketing their weapons, and intensive industry lobbying to get the Export-Import Bank to provide credit for arms trade.[19] It may be hard, but I refuse to believe it is impossible that a public made fully aware of such goings-on cannot find a way to correct them. If the point is to keep employment up and factories open (and anyway, "if we don't sell it to them, the French will"), is there no way in God's name or the Earth's to subsidize competition in and foster credit for the transfer of medical technology, energy-conserving patents, and water-treatment plants?[20]

And finally, I have one last quarrel with the prevailing literature. We have all been so battered with dire conjectures, generally overstated and ordinarily unrelieved by any glimpse of solution, that it tends to numb the public into fatalism—a sense that the earth has a cancer that is untreatable. My aim is to turn the apprehension into a more positive direction. The goal is to identify the truly serious global environmental problems, and, in particular, to determine what underlies them. What should our agenda be? Why is it hard to repair and defend the global environment? And what can be done?

✢ *Acknowledgments* ✢

DURING the course of gathering my thoughts, I had the opportunity to present and hone my ideas before a series of audiences at Amerika Haus (in Berlin, Frankfurt, and Munich), Case Western Reserve University, Columbia University, University of Copenhagen, Dominican College, Duke University, University of Göttingen, University of Hawaii (Spark M. Matsunaga Institute for Peace), Lewis & Clark University, McGill University, Merz Akademie (Stuttgart), New York University, University of North Carolina, North Carolina State, Oxford University, University of Tennessee, University of Toronto, and the University of Southern California. Additional "test runs" were made at colloquia sponsored by the American Philosophical Association (Western Division), the American Bar Association (Section of Natural Resources, Energy and Environmental Law; International Environmental Law Committee), the American Society of International Law, California Bar Association (Subcommittee on International Environmental Law), the Environmental Protection Agency, the Fourth International Conference on Environmental Futures (Budapest), and the Institut für Umweltrecht (Bremen). I am grateful to all those who made these appearances possible, including the U.S. Information Agency; whatever the events did for the audiences, each interchange benefited the author and the manuscript that follows.

In addition, in 1990–1991, in anticipation of the 1992 United Nations Conference on Economics and Development at Rio (the "Earth Summit") I had the opportunity to serve as Reporter for the American Bar Association's Project ABIGAILE (Intersectional Committee to Prepare ABA Resolutions on the International Law of the Environment). While the ideas that follow are hardly the views of the ABA—there were in fact many differences of opinion—I am grateful for the helpful back-and-forth with members of the project, in particular Frank Friedman, Sandy Gaines, and Daniel Barstow Magraw.

Parts of chapter 1 are an expansion of "Preserving and Repairing the Global Environment: The Rapporteur's Draft for the ABA Intersectional Committee to Prepare Proposed Accords on the International Law of the Environment," which was copyrighted by the ABA, published in volume 2 of the *Kansas Journal of Law and Public Policy* (Spring 1992), and appears with the ABA's permission. Other passages, particularly of chapter 8, were published initially in "Beyond Rio: What Should a Climate Change Convention Aim for? 'Insuring' Against Climate Change," in volume 86 of the *American Journal of International Law* (1992), and appear with permission of the American Society of International Law.

Other passages appear or are scheduled to appear in "An Essay in Commemoration of Biosphere Day: Toward a Global Commons Trust Fund," in volume 19 of *Environmental Conservation* (Spring 1992); in "Law, Morals, and the Biosphere," to be published in *Surviving with the Biosphere: Proceedings of the Fourth International Conference on Environmental Future*, edited by Nicholas Polunin and Sir John Burnett (Edinburgh University Press, 1992); in "Healing the Seas through a Global Commons Trust Fund," to be published in *Freedom for the Seas in the 21st Century: Ocean Governance and Environmental Harmony*, edited by Jon Van Dyke, Durwood Zaelke, and Grant Hewison (Island Press, 1993); and in "Überlegungen zur Umweltkrise: Die spirituellen, Juristischen und institutionellen Wurzeln," in volume 3, no. 3, of *Informationdienst Umweltrecht* (July 1992).

Thanks are owing, too, to a sequence of summer law student research assistants who left their mark on the manuscript: Kimberly King, Larry Helfman, Karen Chang, Maureen Edward, Dave Stejkowski, and Eric Claeys. Generous support for the summer research was provided by the Conrad N. Hilton Foundation through the U.S.C. Law Center.

I want to acknowledge some special contributions along the way of Bill Burke (marine resources), David Caron (international law), Howard Chang (law and economics), David Goble (wildlife), Bob Freidheim (law of the sea), Tom Griffith (game theory), Barbara Herman (philosophy), Ed McCaffery (law and economics), Larry

Simon (insurance and law), Gene Skolnikoff (science and policy), Rip Smith (international law), and Jeff Strnad (law and economics). Any slips that survived are my own undoing.

Valuable editing suggestions were provided by Anna Ascher at the *American Journal of International Law*; Angela Conway and Michelle McNulty at the *Kansas Journal of Law and Public Policy*; Alice Calaprice, senior editor at Princeton University Press; and Susan Breyer, Carey Stone, and Ann Stone. For secretarial services, Yolanda Frais is to be thanked; for additional word processing, Shirly Kennedy and Barbara Yost.

Finally, it is a pleasure to honor, once more, the outstanding work of the U.S.C. Law Library under the direction of Albert Brecht. For this book, for a period that covers at lease five years, Pauline Aranas, Paul George, Rob Jones, Hazel Lord, Lisa Moske, Sue McGlamery, Leonette Williams, and Will Vinet adroitly and sunnily tracked down, ferreted out, and accessed publications and raw data on ozone, whales, wolves, Third World debt, forest cover, industry profits, global temperature, geosynchronous orbits, and the budgets of sundry agencies. What a job. This is their book too.

The Gnat Is Older than Man

GLOBAL ENVIRONMENT AND HUMAN AGENDA

✣

Diagnosis: The Earth Has Cancer, and the Cancer Is Man

IN THE DOCTOR'S JUDGMENT

SOME THIRTY years ago in California, a symposium was convened on the state of the earth. Among the many speakers, who included eminent geochemists, botanists, and demographers, was an elderly physician. All that he could contribute, the doctor said almost apologetically, was a medical appraisal of the planet: how he would diagnose the earth were it a patient who had dropped by his office for a checkup.[1]

Typically, the doctor explained, a checkup begins with a count of blood cells. If he should find, for example, that within the space of a month the white cell count had risen from 5,000 to 25,000, he would suspect that something was amiss: quite possibly an early stage of leukemia.

Now, he continued, suppose we were to regard, as well we might, the various features of the planet—the flora, the animal kingdom, and so on—as a federation or community of interdependent organs and tissues that go to make up the patient under review, the world as a whole.

If this planetary patient, sitting in his examining room, could recite its medical history, the most salient fact would be this: that the human component of the earth, its self-conscious tissue, had increased from 500 million in A.D. 1500 to 760 million in 1750 to 2 billion in 1954, the year of his speech. (Today, a mere four decades later, we have passed 5 billion.)

As a doctor he would therefore ask himself this question: "What if it became evident that within a very brief period in the history of the world some one type of its forms of life had increased in number and

3

obviously at the expense of other kinds of life?" To make a long diagnosis short, our doctor said he would judge "*that the world has cancer, and that the cancer cell is man.*"

Among the parallels the doctor drew:

Cancerous growths demand food; but so far as we know, they have never been cured by getting enough. "At the center of new growth, and apparently partly as a result of inadequate circulation, necrosis often sets in—the death and liquidation of cells that have . . . dispensed with order and self-control in their passion to reproduce. . . . How nearly the slums of our great cities resemble the necrosis of tumors."

When a cancer grows, it exerts pressure on adjacent structures and organs. So, too, with humankind and its environment. Just as a brain tumor presses against cranial capacity or, metastasizing, dispatches colonizers to invade other organs at a distance, so humankind engages in mass migrations, even border warfare. "To what degree can colonization . . . be thought of as metastasis of the white race?" "The destruction of forests, the annihilation or near extinction of various animals, and the soil erosion consequent to overgrazing, illustrate the cancer-like effect that man—in mounting numbers and heedless arrogance—has had on other forms of life on what we call 'our' planet."

The Weakness in the Medical Diagnosis

There are some evident objections one can make to the medical analogy of Dr. Gregg (shortly to die of cancer). The interdependency of the earth's parts does not amount to the interdependency of organs within a true organism. Deprive a higher organism, a mammal, for example, of its heart or lungs, and there appear clear, quick signs that the organism has perished. The earth as a whole, including its life web, is not as fragile. Although there is a Gaian view of the earth that emphasizes extensive cooperative relationships among the oceans, ice packs, atmosphere, and living creatures,[2] the Gaian interrelationships are not so finely, so precariously, tuned. However much we might deplore elimination of the world's exotic species of animals or the annihilation of great forests, it is not clear in what sense their excision from the earth would "kill" the planet. Indeed, one teaching of the Gaian literature, ironically, is the *resilience* of the planet. Life has survived more than one bashing by asteroids, as well as a profound revolution in the composition of its atmosphere— from one suffused with methane (which would kill us) to one of 20 percent oxygen. Life, human life in all events, is still here, and—to put it mildly—flourishing.

And even if we regard as our patient not life on earth as it might evolve under exotic circumstances, but life-and-living-as-we-know-it, the role of population pressures in its maladies is not straight-forward. It is truer to say that the environment is being impaired in different ways by different societies. Such problems as desertification and soil erosion, common plagues of the less developed Third World, are indeed directly traceable to population pressures. Desper-ate peoples seize cattle droppings, which should be permitted to sink into and restore the fertility of overtaxed soil, to be burned as fuel. But some of the larger, more widely publicized ills—including the thinning of the ozone shield and the specter of global warming—are largely the product of fuels and chemicals associated with highly industrialized nations, many of which display rather stable, even de-clining birthrates. To put it the other way around, we can agree with the doctor in favoring population controls (which could be advanced by, for example, conditioning international developmental loans on the borrower's adoption of population measures); but we do well to keep in mind that even if population worldwide should be curbed, the health of the planet would still be in jeopardy.

In what ways, specifically?

DIAGNOSIS AT RIO

Suppose we were to carry forward the medical metaphor, and up-date it. The diagnosis today would consist in the list of afflictions that brought the world's leaders together to the United Nations Conference on Environment and Development (UNCED), or "Earth Summit," in Rio de Janeiro in the summer of 1992. The bottom-line judgment is this: almost imperceptibly, but at an accelerating pace, pressures from the swelling of the globe's human population, compounded by the backlashes of industrialization, appear to be blighting the elements on which a healthy earth and decent life depend.

Population experts expect another three billion people by 2025. Ninety-five percent of the increase is projected in the less developed—which is to say, already more struggling—countries.[3] While population is not the only problem, it does raise the question: With so many living at the margins of existence now, how will we and all the newcomers make do tomorrow?

Fresh Water

Start with water. Worldwide, sources of potable and even agricultural-grade water, both above ground in rivers and below ground in aquifers, are suffering from overappropriation and pollution. In Poland, half of the water is reportedly too polluted even for industrial use.[4] In the Middle East, the impending water deficit is inching toward an international crisis potentially more explosive than religion or statehood.[5] The problem goes beyond satisfying thirst. Unsanitary water is the world's most serious health problem. The World Bank reports that one billion people in developing countries have no access to clean water.[6] Every year millions of people—predominantly children under age five—are victims of dirty-water diseases including diarrhea, cholera, and typhoid.[7]

Farmlands

Farmlands are under pressure, too. Rich topsoil is washing away into rivers or, overgrazed and underwatered, blowing away in the wind. Elsewhere, arable land is being lost or degraded through waterlogging, acidification, salinization, and rising water tables sated with toxic chemicals. Where once the word "refugee" brought to mind persons displaced by war, we now have refugees from desertification. In Africa, along the Sahel's 3,000-mile front, there are 10 million wandering victims of a soil gone barren.

Forests

Forests are being stripped, stressed, and burned. In Europe, airborne industrial wastes—acid rain and acid fog—are implicated in a widespread *Waldsterben* (literally, "death of the woods"). In much of the Third World, the reduction in biomass signifies something worse than infringement of "recreational" opportunities. Almost one half of the world's population depends on fuelwood as its principal energy source.[8] Many face a future uncertain as to how they shall cook or keep warm.

In the tropics, deforestation presents yet other threats. The globe-spanning belt of tropical moist forests, now under siege from subsistence farmers and developers, are particularly rich in plant and animal species and constitute a major factor in drawing excess carbon dioxide out of the atmosphere. Once ravaged, these tropical forests are particularly hard to reestablish. "Instead of having a band of greenery around the equator, the earth may eventually feature a bald ring. . . . Ecosystems [that] have been in existence for at least 50 million years are being eliminated within a period of a half a century or so."[9]

The Oceans

The oceans, to which some futurists once looked as the next frontier for feeding the world, are being tampered with as well. The omens are everywhere. Across the world, coral reefs are "bleaching" as life departs them. Increasingly frequent red tides and massive die-offs of marine mammals such as harbor seals and dolphins are early messengers of a similar warning. It is not just chemicals that are to blame. Modern technology seasons the seas with millions of tons of stubbornly persistent litter. Each year, tens of thousands of marine mammals, turtles, seabirds, and fish die from entanglement with or ingestion of plastics and abandoned fishing gear ("ghost nets"), some of which will not disintegrate for centuries.[10]

The same technology has made it possible for the intensive-fishing nations, having depleted favored stocks along their own coasts, to sweep distant seas with huge nylon drift nets 30 miles long, "the single most destructive fishing technology ever devised by man."[11] Commonly, only a fraction of the haul is the commercial catch the drift-netters are seeking. The rest of it—nontarget fish, turtles, marine mammals, sea birds, the living environment of the ocean that had the bad fortune to be in the wrong place at the wrong time—is "wastage" that is simply tossed back overboard to die.[12]

This "strip-mining of the ocean" is not only imperiling the aquatic environment; the new technology represents an escalation in brutality: "Scientists in the last two years have recorded more that 200 dead whales and dolphins showing net markings and with their fins, flippers and tails slashed. Turtles have been found with amputated front legs. The fishermen mutilate the animals, dead or alive, to untangle them from the nylon filaments."[13]

Even if the drift-netting should be brought under control, a return to "ordinary" fishing would not be much comfort: as early as 1979—before drift nets became prevalent—competition among fishing nations had driven the annual world marine catch 15 to 20 million tons (20 to 24 percent) lower than it might otherwise have been without overfishing—and was exterminating 7 million tons of nontarget fish and a million seabirds each year.[14] Increasingly, the fish that does make it to market is tainted. A 1991 *Consumer Reports* survey of supermarkets found that 43 percent of the salmon and 25 percent of the swordfish tested positive for polychlorinated biphenyl (PCBs). Ninety percent of the swordfish also bore mercury.[15]

The worst may be yet to come—and practically unavoidable. At the close of World War II, the Allied forces, harried and with no easy alternatives, dumped hundreds of thousands of Nazi chemical weapons into the shallow Baltic, and then topped it off with an ocean deposit of 300,000 tons of ready-to-fire weapons (principally into the Baltic), as well. There it all sits—until the casings corrode and release the lethal contents to do what it will to the marine and plant life and the 30 million people who live along its coastlines.[16]

The war refuse is not lonely. It has since been joined by canisters of nuclear waste which, dropped into the ocean for "safe storage," are already showing signs of fatigue.[17] Indeed, where nuclear waste is concerned, traditions of military secrecy have compounded the problem. An ex-Soviet official recently admitted that for nearly thirty years the Soviet military had been jettisoning its nuclear wastes (including thousands of canisters, twelve old reactors, and one damaged submarine) into the Arctic Ocean in the most heedless way imaginable. Impatient Soviet sailors got the canisters to sink more quickly by punching holes in them.[18] The canisters, perforated and intact, are still lying there.

Wetlands

The coastal wetlands, the fertile womb of marine life necessary to restore these threatened losses, are vanishing under the pressures of commercial development and a siege of sewage and waste of gargantuan proportions. The deeper reaches of ocean and seabed, less rich in life to begin with, are the ultimate cesspool for all this coastal runoff, as they are for additional loads of airborne pollution that blows out from the land and toxic wastes that are dumped or incinerated directly on the ocean's surface.

Biodiversity

Meanwhile, on land, plant and animal species are being eliminated or placed at risk faster than scientists can sort and inventory them. Some of the slaughter, such as of whales, elephants, pelt and "game" animals, is someone's deliberate action. More commonly, the loss is the unintended consequence of habitats being destroyed by development, or by the careless introduction of an exotic species which, unchecked by any native predator, proceeds to destabilize the local ecological balance.

The consequent reduction in biological diversity is not only an impoverishment of nature, a "cosmic insult," but is also a concern even from humankind's most selfishly self-interested point of view. Even today, medicinal advances remain heavily dependent upon a supply of newly discovered natural ingredients from plants and animals. Both the Pacific yew and the African clawed frog made the

9

front page of the *New York Times* within recent years, the first as the source of a promising cancer medication, the latter of a new, natural antibiotic.[19]

The same, yet-to-be inventoried stock of genetic material is also of value to agriculture. This is because the crops that flourish in any decade are at the risk of being weakened or obliterated by disease, pests, or rapid shifts in climate and soil conditions. In modern times the risk has been increased by the spreading reliance, across vast harvest areas, of a single high-yield strain of each of a few crops (a practice known as "monoculture"). One insurance against the risks of collapse has been to maintain a broad "portfolio" of crop varieties whose genetic codes are held untapped in the wilds. It is this portfolio that we are now liquidating—just when the prospect of a sudden climatic change and the resultant need to call upon such natural genetic "insurance" may be the most critical.

Toxic Waste

Twenty years ago there was a feeling that the earth was going to "run out of" resources such as fossil fuels and basic minerals that are the key factors of an industrial society's production processes. More recently, at least some of those fears of resource shortage have been alleviated by technological substitution: copper being replaced by glass and plastic, for example. The concern today is shifting from limits on what the earth can supply (the input of industrial and agricultural processes) to limits on what the earth can safely absorb (the by-product output). Ocean pollution and ozone depletion are each facets of this looming waste-storage problem. But the question of technological by-products comes up in other forms. Perhaps no threats are more acute than those posed by nuclear power plants, particularly the shabbily built and shoddily managed units in Eastern Europe; like the plant at Chernobyl, they cast a long shadow.[20] And even if the plants and factories *operate* safely, nuclear waste is generated just in the ordinary course of operations. Where to hide it?

Even aside from nuclear waste, the question of storing the world's more lethal garbage—by 1981 an estimated 300 million metric tons[21]—has become a major environmental challenge.* Isolated dumpsites have long been the answer of choice (that is, a better choice than slyly pouring the wastes into storm drains and waterways), but the dumpsites are now the problem. Some indication of the magnitude of the task is suggested by experience in the United States, which has had the resources to get a jump on other countries in cleaning up its sites. After considerable publicity over conditions at Love Canal near Buffalo, New York, the federal government identified over 32,000 hazardous waste sites, placing, thus far, over 1,200 of them on a National Priorities List (NPL) for cleanup. From 1970 to 1990 the work force assigned to the task had grown from 5,550 to 17,170, and budgets increased from $1.3 billion in fiscal year 1971 to $5.1 billion in fiscal year 1990. Yet cleanup had been completed at only fifty-two sites (and begun at but five hundred others).[22]

The problem is of course worldwide, and transcends boundaries. By the 1980s, a huge international traffic in transboundary waste had developed. Typically it is the poor Third World nations that agree, for a fee, to accept for storage under their lands the toxic garbage of the rich; too often, declining prices for the raw materials of the poor countries, and their huge burdens of debt, leave them unfinnicky.[23]

The Ozone Shield

In the stratosphere, the fragile layer of ozone that shields life below from some of the sun's potentially lethal ultraviolet radiation is under assault. When the thinning was originally discerned, it appeared to be localized above the Antarctic and to be essentially a winter phenomenon. But more recently, studies have indicated it is widening across space and time.[24]

---

* This figure does not include the tons of chemical weapons stockpiled around the world which now have to be disposed of safely before they simply leak—a problem that continues to stymie everyone even after enormous investments in studies and in alternative technologies. See U.S. Congress, Office of Technology Assessment, *Disposal of Chemical Weapons: Alternative Technologies* (Washington, D.C.: Government Printing Office, 1992); Stockholm International Peace Research Institute (SIPRI), *Chemical Weapons: Destruction and Conversion* (London: Taylor and Francis, 1980).

11

The evidence is now overwhelming that chlorofluorocarbons (CFCs), halons, and other gaseous by-products of air conditioners, fire extinguishers, computer-chip solvents, and other modern contrivances are implicated in the chemical reactions that are depleting this protective covering. To put the peril in its bleakest light: before the "ozone shield" formed eons ago, life on earth was restricted to sea creatures that lived deep enough in the water to be defended from the mutating power of the rays. It was only with the formation of the shield that life was able to forsake its original briny sanctuary and colonize the land.

No one expects a complete collapse and reverse migration. But the thinner the ozone layer, the more ultraviolet radiation will pass through to the earth's surface, increasing the incidence of skin cancer and cataracts. There is also some concern, yet unsubstantiated, that the radiation may suppress the responsiveness of the human immune system,[25] and it clearly affects the growth of plants and marine life that occupy upper zones of the ocean.[26] Placed at risk are the many fish that spend their larval or juvenile lives near the surface, as well as the phytoplanktons, the immense communities of tiny aquatic plants that form the basis of the ocean food chain (and are an appreciable force in drawing carbon dioxide out of the atmosphere).[27]

Troposphere

Below the stratosphere, there are prognostications no less ominous. Each year human activities belch hundreds of thousands of tons of carbon monoxide, nitrogen oxides, sulfur dioxide, hydrocarbons, and various particulates into the troposphere (the layer of the atmosphere nearest the earth's surface). There they, and the various products of their interaction such as smog and acid rain and fog, suppress crops, undermine human health, and corrode just about anything man can make, from the newest synthetic materials to the most ancient cultural artifacts. Seventy percent of all city dwellers across the world—1.5 billion people—breathe air that health experts consider unhealthy.[28] Even intelligence may be undermined: reports from Bangkok, where traffic congestion is chronic, suggest that children may lose four points of IQ by the age of seven from high levels of lead in their bodies.[29] And the effects are by no means strictly local. The atmospheric currents have become a major conveyer belt for far-flung blighting, shuttling "local" toxic effluents to the seas and lakes, and mantling even the remote Arctic in haze.[30]

Climate Change

The most widely publicized (and probably most complex and controversial) phenomena involve prospective long-term changes in climate variables. The earth is enveloped by a blanket of gases that traps some of the solar radiation that has reflected off the earth's surface and is heading back toward space. These gases (denominated greenhouse gases, or GHGs) include water vapor, carbon dioxide (CO_2), CFCs, methane, nitrogen oxides, and certain trace gases.

Some such trapping of outbound radiation is vital, for without any greenhouse effect the earth would be perhaps 60° F colder than it is—essentially unlivable for familiar life forms. The problem is that human activities, particularly since the industrial revolution, are steeply increasing the concentration of GHGs.

Because of the quantities of GHGs already loaded into the atmosphere, the expected "residence" period of the various GHG molecules (centuries, in some cases), and the inertia of massive societal commitments to GHG-producing activities (such as heavy capital investment in fossil fuel use), experts seem generally to agree that by the middle of the next century—within present lifetimes—the atmospheric concentration of CO_2 (and its "blocking" equivalent in other GHGs) will at least double from preindustrial levels.[31]

It is not certain what will happen if the GHG buildup continues along present trends unabated. *Other things being equal*, there are strong grounds for concern. In theory—a theory that has been uncontested since its proposal in 1896[32]—the thickened blanket should alter the radiation balance, destining the planet to retain a larger proportion of the sun's energy.

All this is fairly uncontroversial. But it is agreed, too, that all things may not be equal, and the facts that have been gathered thus leave many important questions unanswered about the timing, magnitude, and likely consequences—-both physical and social—of global warming.

The risks are these.

The most authoritative international pronouncement, that of the Intergovernmental Panel on Climate Change (IPCC), reports that global mean surface temperature has increased from 0.3° to 0.6° C over the past century.[33] The 1980s produced six of the ten warmest years since records began to be kept in the 1800s; and since then, 1990 and 1991 were reportedly the hottest on record.[34] The IPCC anticipates that if we maintain present trends unchecked (the "business as usual" scenario), we should anticipate a mean temperature in 2025 about 1° C higher than what we are experiencing today; it would be 3° C warmer by the end of the coming century.[35]

Those who find the projections alarming tend to emphasize a historical perspective. The mean global temperature has not fluctuated beyond 5° to 6° C during the past 125,000 years.[36] And oscillations of lesser magnitude have been marked by the coming and going of ice ages and radical shifts in environmental composition. Those oscillations, moreover, were measured in tens of thousands of years. Humankind and other living species face the prospect of such a temperature change over the space of decades.[37]

Most worrisome, the global temperature represents an *average* across the world, one that would not be spread evenly across regions. Some areas, generally supposed to be those closer to the equator, would experience relatively little of the increase, while areas in high latitudes (closer to the poles) would likely experience increases of over twice the mean.[38] If so, a global increase of 3° C might confront some regions with increases of 6° C or more.

The consequences, as well as timing of such temperature shifts (should they come) are themselves subject to uncertainties. But any new level of trapped energy would have to find *some* way to express itself—as by altering ocean currents or intensifying weather phenomena such as windstorms, sea surges, and floods. In the succinct judgment of an insurance analyst: "Put in very simple terms, more energy means more catastrophe."[39] We are not only speaking of hurricanes, inland floods, and mudslides; the list of potential catastrophes includes, ironically, additional frosts and freezing damage in certain areas.[40]

There is the risk, too, that agricultural productivity would be impaired in regions beset by drought, if not by outright desertification, and perhaps assaulted by novel weeds and pests. Pressures on species and habitats, already a subject of concern, would intensify. Forests, for example, would be challenged by new pests and more fires and perhaps would be unable to migrate in pace with shifting weather zones.[41]

By and large, human health has not been advanced as a major issue, but some concern has been expressed over added risks of insect-carried and other viral diseases, which tend to flourish in warmer, wetter environments.[42] Marine productivity could be impaired by loss of wetlands and disturbances in hydrochemical as well as hydrological conditions—for example, increased acidity and shifts in upwelling (feeding) zones. Energy supplies would have to meet increased demand for air conditioning and water transport and delivery.

Far more worrisome is that a rising sea level would threaten human settlements. If the earth warms, some elevation in sea level can be expected from thermal expansion—the warming of ocean waters—alone. It has been estimated that a warming of between 1° and 2.6° C could result in a thermal expansion contribution to sea level of between 12 and 26 centimeters by 2050.[43] And there is the further risk that the heating will trigger a partial melting of alpine and even (a more remote prospect) polar ice packs. Although the glaciers that once covered much of the Northern Hemisphere have retreated, the world's remaining ice cover contains enough water to raise sea level over 75 meters.[44]

In recent years some of the more extravagant inundation scenarios have been scaled back, but the IPCC predicts a rise of 20 centimeters by 2030, 30 to 50 centimeters by 2050, and 65 centimeters to 1 meter by the end of the twenty-first century.[45] In view of the fact that one third of the world's population lives within 40 miles of the sea, even a rise of 20 to 30 centimeters would cause considerable damage, most ominously in the large, densely populated deltaic areas of

Egypt and Bangladesh.[46] Small island nations, such as the Maldives and Vanuatu, could submerge entirely.

Moreover, some authorities have suggested that as the surface of the ocean is heated, storm phenomena such as cyclones and hurricanes, which are generally believed to debut when the ocean surface exceeds 26° C, could be expected to increase in incidence and intensify, exacerbating the damage to coastal communities and ecosystems that presumably will already have been made more vulnerable through the rising sea level.[47] Wetlands would be at risk, freshwater sources invaded, and multibillion dollar investments in sewage and storm systems impaired.

Arctic and Antarctic

The Arctic and Antarctic, once considered to be the ultimately "safe" pristine frontiers, are swiftly soiling. Scientists were recently amazed to discover that as the waste of the industrialized Northern Hemisphere is swept poleward by winds, currents, and animal migration, the native population, the Inuit, have developed higher levels of PCBs in their blood than any known population on earth that has not been a victim of industrial accidents.[48] In both polar regions, but particularly in the Antarctic, the prospects of oil and mineral exploitation—although presently under an uneasy, temporary suspense—bode worse degradation to come. Indeed, if one is to gather together the most troubling possibilities, we might remember that three fourths of the world's fresh water lies locked in the great ice caps of Antarctica, the legal ownership of which—10 percent of the earth's land surface—is presently unsettled. Will the Antarctic, always under threat of being trampled for its mineral wealth, become a battleground for the water-starved?

In considering the polar regions, one must also account for the fact that threats to the scale of the earth's ice cover have been associated with worrisome dynamics unrelated to sea level rise. The ice-covered surface of the earth has an extremely high albedo, that is, reflecting capacity. Presently the ice caps serve to "mirror" back to space most of the sunlight that strikes them with minimal absorption of heat. To the extent that ice-covered areas recede, the earth that is exposed in their stead will present to the incoming energy a darker and therefore more heat-absorbent face. Also, under the frozen Arctic permafrost areas, vast stores of methane are presently locked up in the form of clathrates—molecular lattices of methane and water. If the warming releases them, they will migrate to the atmosphere and magnify the warming even further.[49]

The Environment in War

Recent developments in military technology have added another direction from which the environment is threatened: it may become a casualty of war. Such a specter has been raised in connection with the "nuclear fall" that many scientists predict would follow an extensive nuclear exchange.[50] But even hostilities on a more limited scale threaten to draw the environment into warfare in disturbing new ways. The war in Southeast Asia demonstrated that the arsenals of advanced nations already included, by the 1970s, such environmental-modification techniques as defoliation, rainmaking, and massive plowing. During the Persian Gulf War, allegations were made of deliberate oil discharges into the gulf to advance military (and perhaps political) objectives. The Gulf War also aroused speculation about the possible strategic benefits—and dangers—of igniting oil fields and thereby initiating modifications in local and regional weather patterns. Many more sophisticated enlistments of the environment in wartime have been publicly discussed, while still others have undoubtedly been reviewed by military strategists, in private.[51]

The Geopolitical Dimensions

Finally, as if there were not enough work just straightening out relations with our besieged environment, we cannot fail to note that the environmental pressures are fraught with the potential to pit human against human. In the ancient world, prolonged and severe climate change led to uprooting of whole populations. But today, populations are denser and mass migrations would transgress political boundaries and exacerbate cultural tensions—with all the ominous

17

frictions such conflicts portend. What respect for any basic notions of property, boundaries, and civil order shall we expect to survive, if environmental degradation were to impose increasing stresses on food and water supplies, if arable zones were to shift and traditional population centers were to sink beneath rising tides?[52]

Five of the world's fourteen major river systems are shared by four or more nations. Just to suggest the potential explosiveness, in the already tense Middle East, the Jordan River basin, on which Israel, Jordan, and the Palestinians each rely, will almost certainly be inadequate to support the projected population of the region (it may be at 15 percent deficit) by the year 2000. A much (and perennially) exaggerated "oil shortage" occupies the most media attention, but those on the forefront of pondering world resources are already talking in terms of emerging "water powers," most notably Turkey, which has the strategic upper hand on the flow of the Euphrates into Iraq and Syria—a far more incendiary blessing than oil. In fact, months before the outbreak of the Persian Gulf War, Syria and Iraq had threatened hostilities against Turkey if Turkey did not cease diverting water for impounding in its new $21 billion Ataturk Dam irrigation project.[53] The Turks claim that Syria is sponsoring Kurdish terrorism as a way of maintaining leverage in water talks.[54] And it should not be forgotten that in the early 1960s, when Syria still held the Golan region, it tried to build a canal to channel water back to Syria; the Israelis bombed the project as a precursor to the 1967 war.[55]

Weather control would be one antidote. But even as humankind—or, one might better say, the advanced nations—acquire the capacity to modify the weather, it will likely prove a mixed blessing. Edward Teller, the "father of the H-bomb," once suggested that conflict over weather control would be the most likely cause of "the last war on earth."[56]

A SECOND, LESS FOREBODING OPINION

Not everyone concurs in so dismal a diagnosis.[57] There are hopeful signs. Sulfur dioxide emissions have been reduced, at least among the major industrial nations.[58] By global agreement, CFCs are being eliminated—a striking testament to the potential of diplomacy. Almost without public fanfare, the growth rate in both CO_2 and methane emissions began to show a hopeful decline even before the attention GHG reductions received at Rio.[59] By separate agreements, the nations bordering the northeast Atlantic Ocean have agreed to stop dumping munitions in that sea beginning in 1993, and to impose a fifteen-year moratorium on the deposit of nulcear waste effective in 1992. With the spread of birth control devices, some areas of the earth are getting a grip on population: in Brazil, between 1970 and 1990 the average number of children per woman of reproductive age fell from 5.8 to 3.3.[60] There are more trees in the United States today than there were in the 1920s. The reasons are instructive: "Oil and gas replaced fireplaces and stoves that used to burn 400 cubic feet of wood per American per year; cars, trucks and tractors replaced the nation's fifty million horses and the one-third of all agricultural lands that fed them; advances in seeds and fertilizers mean the nation's crops are being grown on about 400 million acres, the same amount under cultivation in 1920. On the rest, trees grew again."[61]

This last illustration suggests another consideration. The most apocalyptic predictions call for projections over half a century in the future. The history of doomsaying is, fortunately, replete with failed forecasts over more modest time horizons. Only two decades ago, Paul R. and Anne H. Ehrlich, whose books, including *Population Bomb* (1968) and *The End of Affluence* (1974), deserve credit for having focused popular attention on population and resources, brimmed with predictions of massive catastrophes, principally famine-connected. In 1968 the Ehrlichs were notifying us that "the battle to feed all of humanity is over. In the 1970's . . . hundreds of millions of people are going to starve to death in spite of any crash program embarked upon now." In 1974 we were being prepared for "a nutritional disaster that seems likely to overtake humanity in the 1970s (or, at the latest, the 1980s)," a situation "that could lead to *a billion or more people starving to death.*"[62]

The eighties have now passed; and what we find is not billions starving, but a food supply that has been continuously improving, as measured by grain prices, production per consumer, and the famine/death rate.[63] In fact, from the mid-1960s at least to the mid-1980s, world food production appears to have increased faster than the rate of world population growth, which has been decelerating.[64] Dreadful malnutrition exists in certain areas, but the cause often stems not so much from a deficit of procurable global supplies as from the collapse of human relations, most notoriously, civil wars, some of which have sunk to the deliberate obstruction of relief efforts.

The Other Side of the Climate Change Alarm

Indeed, as for climate and global temperature, it is at the least ironic, in reviewing the literature of the 1970s, to be reminded that the peril the Ehrlichs were broadcasting then was not greenhouse warming, but a crop-ravaging global chill.[65] Similarly, Stephen Schneider, the climatologist author of *Global Warming* (1989), was in 1976 relaying "the warnings of several well-known climatologists that a cooling trend has set in—perhaps one akin to the Little Ice Age."[66]

Those who oppose immediate massive spending to combat greenhouse warming do not rest their case solely on such ironies, however.

IS THE EARTH REALLY WARMING?

Even the most basic elements underlying the alarmist's case are hardly clear. Take what one would imagine to be the deceptively simple question: *Is* the earth really warming? We just cannot be certain. While the IPCC claims to discern a warming of 0.3° to 0.6° C over the past century,[67] and the more anxious wing of the scientific community points to the recent spate of "hottest ever" years, critics maintain that the apparent recent rise can be as well explained by an increasing urban heat island effect. Most temperature devices are located near large cities, where readings are distorted by concrete, roads, and other incidents of concentrated human habitation. When data taken at some remove from cities are analyzed, the apparent increase is not detectable. Ocean temperatures have not changed from 1850 to 1987.[68] TIROS-N weather satellites, measuring global temperatures from the vantage point of space, have detected no evidence of a long-term global warming or cooling trend.[69]

There are, moreover, anomalies that cast some doubts on the alarm-generating theories and models themselves. Most of the contended temperature increase occurred in the first half of the century, when fossil burning and GHG concentrations were at lower levels; it has been somewhat more difficult to claim any statistically significant warming in the second half, when the cumulated loadings have obviously increased. Finally, all the theories predict that warming should be magnified at the polar regions. Yet all agree that temperatures in the Arctic, at least, have declined.[70]

IF THE EARTH IS WARMING, ARE GREENHOUSE GASES RESPONSIBLE?

Second, even if it could be shown that the earth is on a warming trend, it would be a further leap to pin the blame on human activities. The most influential factors by far are incoming solar radiation, which is subject to fluctuations in solar activity, and cyclical variations in the position of the earth with respect to the sun (the tilt of the earth's axis with respect to the ecliptic), and the eccentricity of the earth's orbit.[71] Even the IPCC acknowledges that the size of the warming it believes to have identified (0.3° to 0.6° C) cannot be excluded from the range of natural climate variability.

WOULD A WARMING WITHIN THE RANGE PREDICTED BE BAD?

Those persuaded that global warming is real, and worrisome, can rejoin with several intriguing explanations why the evidence may have been so scant—or perhaps late in arriving. One is that the man-made warming trend is colliding with the natural glacial cycle. Another is that the oceans have been absorbing a hefty share of heat and CO_2, but that the storage capacity is becoming filled. Another—for which there seems to be considerable evidence—is that some of humankind's dirty activities appear to be canceling out, or at least staving off, others. The burning of dirty coal, forests, and other biomass has formed around the earth a "parasol" of particles, principally sulfate aerosols and smoke dust, which reflects away some of the radiation (which the GHGs from the same burning would otherwise

21

help trap).[72] And a recent United Nations Environmental Programme (UNEP) study indicates that the ozone-depleting agents (being eliminated under the CFC protocols) may be counteracting greenhouse warming by impairing the heat-retention capacity of the lower stratosphere.[73]

But even if the warming trend is real, would it be so bad? On the face of it, even the upper-range projections for average global increases represent a more modest differential than we commonly experience now in a single day flying from Montreal to Los Angeles. How much we should worry depends considerably on the distribution of the average—between night and day, between seasons, and between regions. And with that in mind, a close look at the models is not so disturbing. To begin with, we cannot predict the distribution in any heat rise between night- and daytime. Virtually all of the strong warming is presently projected either for twilight or night,[74] when its effects are imagined to be particularly benign, if not positive.[75] (In fact, this day-night imbalance is consistent with observations in the United States to date.) Moreover, most of the warming is concentrated in higher latitudes, so that popular representations based on the distortive Mercator projections may be rather misleading. In terms of actual areas of the globe, the area subject to greater than 4° C warming is less than 5 percent—predominantly in polar and near-polar areas that are hardly well inhabited.[76]

But isn't exactly that the fear—that the concentrated warming in high latitudes will melt the ice caps, thereby inundating land masses? Here the response turns out to be counterintuitive; the better betting seems to be that global warming would *thicken* the ice caps and tend to lower sea levels. To understand, one has to appreciate the fact that the polar regions are so cold that an increase of even 10° or 20° C is not about to melt the ice sheets, anyway. A. Yanshin, the Russian academician who has been watching the Western hubbub from the sidelines, has been trying to remind anyone who will listen that the thick glacial shield of the Antarctic appears to have formed 30 million years ago and withstood several epochs of climatic warming well beyond the upper range of the pre-dicted greenhouse effect.[77] At the same time, however, a globally warmed environment would experience more evaporation and pre-cipitation. Part of this hydrologic process involves removing water from the sea and depositing it, stowed as snow, on the polar ice packs—thereby thickening them. Recent reestimates suggest that the added glacial mass would have the effect of *reducing* the sea level from 10 to 50 centimeters.[78]

Much the same sort of evidence and reasoning is available to meet the alarmists step by step. Yes, warmer temperatures would increase cooling costs in Tampa. On the other hand, there would be lower space-heating costs—less energy demand—in Boston and Moscow. There is no reason to suppose that the one effect would dominate the other.

Similarly with the fears of droughts and loss of farmland. Yes, some areas would presumably suffer. On balance, though, the amount of precipitation worldwide would be expected to increase.[79] In fact, a warmer, moister, more carbon dioxide-rich atmosphere, together with longer (frost-free) growing seasons, is viewed as gener-ally favorable to the growth of biomass overall (with some variable impacts on plants depending on species and local conditions). The International Council of Scientific Unions' Scientific Committee on Problems of the Environment (SCOPE) concluded that "given the un-certainties in regional scale estimates . . . and in the numerous defi-ciencies in methodologies . . . there is presently no firm evidence for believing that the net effects of higher CO_2 and climatic changes on agriculture in any specific region of the world will be adverse rather than beneficial. . . . It is certain that some will gain and others will lose, although we know neither where they will be found nor the magnitude of the impacts."[80]

In addition, it is true that any interference with existing ocean patterns has to be regarded with misgiving. Yet a leading oceano-grapher has suggested that a warmer ocean, with intensified upwell-ing zones, would make the oceans overall more productive rather than less.[81]

23

Gaps in Our Knowledge

Those who find the future less menacing point also to striking gaps in our knowledge about other aspects of the world, gaps that have to be filled before any dire prognosis can be pronounced. We are only beginning to learn how the world works. Our ignorance is not only about the dynamics of globe-spanning climate and current. Scientists have only started to inventory the world's forests and monitor the thickness of the ice caps. It is common to talk about the oceans being "overfished," but in truth we have only an imperfect idea about the number of fish there are in the vast deep, about how they interact, about how pollution is affecting their environment, or about the maximum sustainable yield for most stocks. As for biodiversity, we do not know how many species there are to imperil. Estimates range from 5 to 30 million—quite a substantial difference, and no scanty figure in either event. Some of the problems are definitional: What is a species? And what is endangered? In the United States each of two thousand "runs" of salmon is designated a separate endangered species—separate on the grounds that no run interbreeds with another. Even population estimating, upon which so much planning seems to depend, has hardly reached the stage of incontrovertible science. If current estimates about future populations are anywhere near the ballpark, we will be the first generation to have guessed it right.[82]

Self-Corrective Mechanisms

Mankind's unintentional self-protective "successes" aside (for example, the cooling of the earth through airborne trash), there is comfort in the sometimes astonishing capacity of the earth to "self-correct"— to maintain, even in the face of considerable disruptive pressures, stability in the conditions vital, even congenial, to life. (That is the thesis of James Lovelock's provocative little book, *Gaia*).[83] Suppose, for example, that we do experience an elevation of greenhouse gases, principally CO_2, and that as a result the earth begins to warm. The consequently warmer climate, plus the abundance of plant-nourishing CO_2, can be expected to invigorate the growth of vegetation even in regions presently inhospitable to foliage, and the new stock, "feeding" on and thereby storing the enriched level of carbon, would operate in the direction of restoring the original balance.

Along these lines, there is much to be learned about the dynamics of cloud formation. Conceivably, as the weather warms, the effect of high-level clouds, which blanket exiting heat and thereby have a warming effect, will be offset by the formation of low-level, moisture-laden marine clouds, which deflect inbound energy and thus exercise a cooling influence. And no one can yet predict the capacity of the sea to absorb rising levels of carbon dioxide and of heat, a process which, if it did not mitigate, could at least defer greenhouse warming by centuries.[84]

Technological Fixes

Such self-corrective mechanisms will certainly expect some boost from mankind's own technological ingenuity. The phasing out of carbon-based fuels and their replacement with environmentally benign substitutes would make a considerable difference in the timing and scale of climate change. Water shortfalls would be ameliorated by a breakthrough in desalinization, and, indeed, a more prudent and widespread deployment of currently available microirrigation techniques. Advances in biotechnology will almost certainly provide some benefits in pest-resistant, water-frugal, climate-adaptable crops. Indeed, there is hope that plants other than legumes might be engineered so as to fix nitrogen in the soil, thereby relieving the need for fertilizers. On an even more visionary scale, experiments are currently under way to increase the ocean's absorption of excess atmospheric CO_2 by artificially nourishing the growth of oceanic algae colonies with seedings of iron.[85]

Bolder (and somewhat shakier) proposals for medicating the globe's afflictions include using laser beams to destroy CFCs before they reach the highest reaches of the atmosphere. It has been suggested that warming could be thwarted by ferrying reflectors into space, or covering much of the world's oceans with styrofoam chips to reflect the light away, or even blocking sunlight with satellites of giant films.[86]

25

Prognosis: Social Choice in the
Face of Uncertainty

Considering the poor track record of the doomsayers and the several reasons to be hopeful, it is natural to ask whether we should just conduct business as usual, taking care of first things—always the nearest term and most evident problems—first?

The answer is clearly no. There are several reasons why complacency about environmental perils, including long-term ones, would be a bad mistake.

Getting the Facts

First, it is true there is a lot we do not know. But gathering the facts and building the models should be one of the principal items on the environmental movement's agenda. Constructing computer models, measuring seasonal changes in ozone and their effects, monitoring gas emissions, and inventorying trees and plant and animal stocks are long-term efforts that require massive financial and institutional support—the sooner begun the better.

Risk Multipliers

Second, the optimists can cite many natural mechanisms that are likely to counteract or mask some future peril, such as the prospect that excess atmospheric carbon dioxide will invigorate the growth of trees, which are CO_2 "consumers." But for every self-correcting (negative feedback) mechanism that is advanced as a possible deliverance from some peril, there seems to be another climatologist coming forward with a positive feedback that threatens to intensify or accelerate the same threat. For example, some suggest that global warming will accelerate the microbial decomposition of the vast stores of dead organic matter that lies on forest floors, thereby speeding up transfer into the atmosphere of vast quantities of carbon dioxide and methane.[87] Another worry is that we simply cannot account for 1 to 2 billion of the net 7 billion tons of carbon dioxide that we add to the atmosphere annually. Even after fifteen years of looking, scientists have been unable to find it either in the atmosphere or the sea. This raises "the unnerving possibility that whatever processes are removing it may soon fall down on the job without warning, accelerating any warming."[88]

Clear and Present Dangers

Third, let us put aside sci-fi scary predictions about the polar ice caps collapsing in 2100 and even bracket, for a moment at least, the whole global warming scenario, which now rivets world attention and fills the environmental groups' coffers but may well prove overstated. We should not lose sight of the clear and present dangers that need to be confronted right now. We know that right now hundreds of thousands of tons of wastes and weapons and worse are lying on ocean floors waiting to corrode and escape. They are not going to disappear. A thinned ozone shield may or may not result in 200,000 additional cancer deaths in the United States by 2040, but every year, right now, millions, especially children, are doomed by easily inoculable and treatable diseases linked to unhealthy environmental conditions. We know that industrial plants presently operating can set adrift murderous clouds of radiation or toxic substances. We know that the oil tankers now sailing our oceans can split open and devastate tidal areas. We know that present-day chemicals, released into waterways or borne by air currents into the seas, are perturbing the marine environment and disturbing the food chain. We know that the expanding scope of global travel has given viruses, such as that responsible for AIDS, a terrible new velocity and range. These are problems that demand international consideration immediately.

Nonlinear Surprises

Fourth, one ought not to dismiss the prospect of some genuine eco-disaster. It is true that each of the most dire scenarios, such as a melting of the antarctic ice caps, might be assigned a low probability when considered independently. But we cannot exclude the possibility that some of the trends will behave in nonlinear ways, increasing gradually for decades, and then suddenly accelerating into a true catastrophe. Nor can we exclude as a real possibility that

27

some presently unforeseen combination of moderate trends, such as a moderate greenhouse warming plus a moderate ozone breakdown plus a moderate ocean acidification, will reinforce to produce a chaotic reaction with genuinely critical worldwide implications not presently foreseeable. One of the odder-sounding, but quite seriously advanced and even riveting scenarios is that the climate change conditions that are being used to forecast global warming will actually—largely through tampering with major ocean currents—trigger a catastrophic cooling.[89] These problems are so complex and potentially of such magnitude that long lead times and considerable marshalling of resources are required just to analyze, much less to respond to them. The sooner we place them on governmental agendas, the better.

Indeed, even if our best predictions are off the mark, public reactions based on them may still be valuable. Suppose we lay a worldwide institutional groundwork for dealing with AIDS, but that AIDS is eliminated before the institution is fully in place. The effort would hardly be wasted. Even though we cannot predict with a high degree of detail when and where the next global epidemic will strike, or say much about its velocity and pathways and character, just establishing and fortifying the right links among national and international health services, such as the World Health Organization (WHO), can do much to shorten the lead time in responding to and containing whatever diseases do arise.[90] The point is, our inability to foresee the exact perils in a precise level of detail should not of itself (absent high costs) discredit present efforts to plan for them as best we can.*

* It is true that the media may overreact to engrossing perils. But it is no less true that politicians, facing reelection on two- and four-year cycles, may systematically underreact to hazards that may not occur for decades. The way politicians discount remote perils could be defended on the grounds that they are presumptively reflecting the attitudes of their electors, for whom the rewards of tax cuts today dominates catastrophe losses tomorrow. Perhaps. But there is an extensive literature indicating how normal people are commonly inclined to misperceive relative risks, and how agencies, given prevailing incentives, are inclined to lay on distortions of their own. See Clayton P. Gillette and James E. Krier, "Risk, Courts, and Agencies," *University of Pennsylvania Law Review* 138 (1990): 1027–1109. The more prudent tack would be to fortify and expand the capacities of our best risk-assessment institutions, such as the Office of Technology Assessment.

Inertia

Then, too, while we do not know exactly what mischief present practices may engender in the future, the effects of what we have done and are doing may acquire a practically irreversible momentum before we appreciate the harm. For example, many of the gases we are releasing into the atmosphere have an expected "residence" of over one hundred years. This is particularly worrisome if we consider demography's own inertial trends. To continue with CO_2, even if the industrialized nations acted immediately to cut their emissions of CO_2 in half (from 1,800 kilograms per person to 900), the projected economic development in the LDCs will almost certainly increase their per capita consumption (from 450 kilograms per person). If some projected rates of population growth in the LDCs holds, a doubling of per capita emissions there would result in worldwide emissions two and a half times what they are today, even if industry were to cut its usage in half.*[91]

Technological Fixes and Flaws

There are, as we saw, those who put their faith in the likely development of technical countermeasures that will enable us to adapt to problems as they become better defined. Rather than prevent against a whole range of conjectural perils, some of which will never eventuate, technical adaptation adopts a wait-and-see mood that promises greater efficiency. But there are several soft spots in the technological optimist position—and not only the obvious risk that we will be waiting and watching until it is too late to correct our course.

It is important to remember that the history of technological solutions is filled with cases in which the technological cure was worse than the disease that motivated it. Chlorofluorocarbons may provide a good example. They were introduced in the 1930s in response to

* Note, however, that China is unusual among major LDCs in having huge coal deposits; much of the Third World lacks its own fossil resources, and it is not clear that, given their poverty, these nations will be able to get their hands on the quantities of carbon that the calculations in the text assume.

29

the hazards of using ammonia as a refrigerant. Ammonia is toxic and flammable. CFCs were nontoxic and nonflammable, as well as cheap and easy to make. (Their discoverer put some of the health-hazard doubts to rest by inhaling CFC vapors and then exhaling to extinguish a candle.)[92] CFCs were so stable, so resistant to combining with anything else, so apparently benign a wonder product, that gradually more and more uses for them were found: as a blowing agent in foam insulation and fire extinguishers, as a solvent for cleaning, and so on. But it was that very stability which has enabled them to migrate intact up to the stratosphere, where we now discover that under the bombardment of high-altitude radiation they give up their chlorine, which in turn attacks ozone—a reaction that no one could have foreseen in the thirties.

Then, too, those who put their faith in technology must realize that technological advances do not occur "of their own." This is particularly true of innovations that commercial markets will not reward. The inventor of a new compact disc player can expect remuneration from buyers. But, as beneficial as he or she would be for humankind, there is no assured commercial market waiting to reward the inventor of a technique for purging the atmosphere of chlorine.[93] A high proportion of funding and other incentives will have to come from a mobilized public sector. And this is true generally: the more massive and expensive and large-scale the technical fix, the more heavily it will have to lean on an express political commitment to get it off the ground, especially at the research and development level.

The necessity that mothers invention can be a legal "necessity" no less than one impelled by some abstract "market." Here, too, CFCs will illustrate. The recent development of substitutes for CFCs in many applications was a direct result of worldwide legal and diplomatic activity, in particular the agreements to curtail the use of ozone-depleting agents. Hence there is nothing inconsistent about having faith in technology and also committing to a fairly high level of global environmental activity—some of which is aimed at promoting the required technological countermeasures.[94]

The Fallacy of "Merely Regional" Crises

It is true that most of the eco-doom scenarios are likely to result in disruptions on at most a regional, rather than global, scale. In this vein, it can be conceded to the optimists, if only for purposes of argument, that worldwide there will be enough food and water and land to accommodate a rising global population. It can be said that foreseeable climate changes are less likely to reduce world production than to shift regions of productivity, so that, for example, some grainlands will move north from the United States and Central Europe to Canada and the Commonwealth of Independent States. The simple picture is one of some winners and some losers but with adequate supplies overall, the only task being to arrange conditions of trade and transport from areas of surplus to areas of deficit.

But looking at the situation from this grand global perspective underrates the problems of redistribution in a world of nation-states, many already desperate, and trivializes the plights of local areas, which alone define the realities that people really face. After all, no individual faces (except perhaps in an almanac) the *world*'s resources and the *world*'s perils, *its* plenitude and shortages. To Africans living in the Sahel, the pains of desertification are not mollified by the thought that there is plenty of water in Russia. Many nations will be able to cope, with dikes and levies, with prospective rises in sea level. But that will not change the fate for vast populations living in Bangladesh and Egypt. One can say of all these problems that they are, in some sense, *regional*; and yet we should remember that the world is too much bound together to regard any catastrophe as "merely regional."

Conclusion

The earth faces a number of perils. Some are immediate. Others, less well-defined, are surely coming. Still others—the most menacing, perhaps—are fairly remote. Almost certainly, some of the more remote and menacing worries will prove to be exaggerated. With equal certainty we can say that other hazards will emerge that no one now anticipates. The aim of this book, in all events, is not to retail horror stories or to arbitrate between the doomsayers and the optimists about what is going to happen if we do nothing in the long term. For the most part we do well to assume, on the side of prudence, that there are real global-scale risks, and to prepare for them within reasonable levels of expenditure.

As two of our leading atmospheric chemists have put it:

> What is particularly troubling is the possibility of unwelcome surprises, as human activities continue to tax an atmosphere whose inner-workings and interactions with organisms and non-living materials are incompletely understood. The Antarctic ozone hole is a particularly ominous example of the surprises that may be lurking ahead. Its unexpected severity has demonstrated beyond doubt that the atmosphere can be exquisitely sensitive to what seemed to be small chemical perturbations and that the manifestations of such perturbations can arise much faster than even the most astute scientist could expect.[95]

Part of our preparation for such surprises is entrusted to science: to gather more facts and improve dynamic models.

And part of the preparation requires philosophical exchange: What sacrifices are merited—what "premium" are we obliged to pay—in order to insure future generations against remote calamities? What value do we place on the environment in its own right? And how ought the financial burdens of medicating the environment to be apportioned between rich nations and poor?

And finally, part of the effort—a central concern of this book—involves the legal and diplomatic structure in which global environmental problems are handled. What reforms need we call for in domestic and international institutions?

CHAPTER II

The Condition of the Earth
from the Legal Perspective

ASTRONAUT AND DIPLOMAT

WHY DO planetwide environmental problems arise, and why do they persist? There are many explanations. Some commentators stress the surge of human population, with the consequent pressures on land, fuel, and other resources. Some point to two-edged "advances" in technology, for whose benefits we pay a Faustian price in hazardous wastes and nonbiodegradable trash. Others pin our predicament on capitalism, or on defects of the human spirit that spring forth in rapacity or a pathological need to dominate Nature.[1] Still others claim the world would do all right were it not for the ineptness and belligerence of local governments, which have been slow to adopt available technologies and to reform administrative practices such as land and water policies. There may be some element of truth in all of those claims—as we shall see.

But granted the manifold origins of our earthly plight, the most important focus should be on the cure. Given the growing worldwide concern that we are veering toward environmental shoals, why do our social institutions have such a hard time correcting our course?

There is a powerful if prosaic place for an understanding of our planetary predicament to begin: from the perspective of an astronaut, at a distance from which the political boundaries, the pointed message of traditional schoolroom maps, are indiscernible. From that remove, what strikes the imagination is the marvelous wholeness of the planet and the globe-spanning activities that connect and sustain its tenants. There is one great envelope of atmospheric gases, the vast body of ocean, the collusive currents of air and water, the broad belts of photosynthesizing vegetation, the complex of plants and micro-organisms that unite in pumping various elements in and out of the atmosphere, all on vast regional and worldwide scales. From space, everything that dominates the attention is unified and interconnected.

Statespersons and lawyers operate from a different, a more mundane, vista. They inherit a world in which all this unity and grand scale have been disrupted into political territories. We all know that most of these penciled borders have little to do with the great natural processes that the geochemist is drawn to, that they fluctuate, that they are often the legacies of chance, intrigue, vanity, avarice, and military battles that could have gone either way. But for all their caprice and impermanence, the boundaries that mark the diplomats' world—hardened, as they commonly are, by pronounced cultural, religious, and socioeconomic differences—are no less to be reckoned with than carbon.

Broadly speaking (for we will have more detail to provide below), the diplomats' maps map two sorts of regions: those that fall under national sovereignty, *national territory*, and those that lie outside the political reach of any nation state, the *global commons*.

The reach over which a nation's territorial sovereignty extends, and the powers and immunities that sovereign nations enjoy, both remain large and fluid topics. It is generally agreed, however, that the area of territorial sovereignty is co-extensive with the nation's geographic borders, extends upward through its air traffic space and, in the case of coastal nations such as the United States, may extend outward through an Exclusive Economic Zone (EEZ), 200 nautical miles seaward from its coast.[2] In this seaward extension, the coastal state has the exclusive rights to fish, to mine and draw oil from the seabed and subsurface, and to police the waters for potential violators of its customs laws and other regulations (including pollution rules).

The global commons refers to those portions of the earth and its surrounding space that lie above and beyond the recognized territorial claims of any nation. At present, that includes the high seas, together with their potentially valuable beds and subsurfaces, that have yet to be "enclosed" by any coastal state as part of its territorial extension.[3] On some accounts, much the same commons status does or should apply to the resource-rich Antarctic, which comprises 10 percent of the planet's land mass. While seven nations—Argentina, Australia, Chile, France, New Zealand, Africa, Norway, and the United Kingdom—have claimed territorial sovereignty over various portions of the continent, at present all the claims are suspended in limbo while discussions continue regarding a number of Antarctic agenda items. These include the settlement of overlapping claims, proposed environmental restrictions, and even whether the whole area should be put under some form of international-community control, perhaps as a World Park, immune from resource exploitation.*

The magnitude and significance of these unowned portions is considerable. At present, perhaps as much as 70 percent of the earth's surface is either commons or not yet subject to undisputed sovereign control.

Moreover, the global commons is not just a surface phenomenon. Above the surface, more exactly, above sovereign air space,** everything is commons. That includes the atmosphere, which, as we have observed, is increasingly being appropriated as a cost-free deposit

---

* In addition, there is a more novel and complex argument that species, or perhaps genetic information, should be declared part of the Common Heritage of Mankind (CHM), like the high seas. See Cyril de Klemme, "Conservation of Species: The Need for a New Approach," *Environmental Law & Policy* 9 (1982): 117. The implications of such a designation would be slightly different for genetic material, however. To declare the high seas and space to be part of the CHM is taken to mean they are at the least open to all users free of charge; there is an alternative, stronger sense to which the United States has never acceded: that designating them CHM means they are the common property of all nations and therefore their value, for example, extractable wealth like seabed minerals, has to be *divided* among nations, whoever does the extracting. In either view of CHM, calls to add biodiversity to common heritage is a threat to the biologically rich countries, many of which, such as Brazil and Colombia, are otherwise poor and inclined to resist loss of control over their own "internal" species and possibly patentable genetic resources. See chapter 9.

** Sovereign air space is itself not a very exact concept, being generally supposed to extend to the altitude achievable by ordinary manned flight, an altitude that continually rises with aviation technology. For a general discussion, see U.N. Doc. A/AC. 105/C. 2/7 (1970).

site for pollutants—with sobering implications for human health, the climate, and ozone shield. Beyond that lie outer space and the planets which, as with the oceans and Antarctic, present "riches" to be squabbled over. Exploitation of the moon and other celestial bodies remains beyond present reach. But, nearer to home, technology has already opened divisive issues of ownership and control over rights to "park" broadcast satellites and spacecraft stations in the choice orbital positions, and over the assignment of broadcast frequencies.*

This division of the planet into regions of separate political significance has enormous significance for the global environment. From any individual nation's point of view, the cheapest solution to *its* airborne pollution is often to build smokestacks so tall that the noxious fumes waft beyond its own territory and across some neighboring nation or out to sea. Tall British stacks have long been a scourge to Sweden; likewise, those of the United States to acid-rained-upon Canada. Liquid wastes are all too often treated (that is to say *not treated*) similarly: they are merely piped into local streams or other waterways, to track a poisonous path down to the open sea.

Given the prevailing framework of international law, the legal and diplomatic response available depends to a considerable extent upon the location of the hazarding actions and of the threatened injury, with reference to this twofold division into national territory and commons areas.

Specifically, we may begin by identifying five possible combinations of (1) where the harm-threatening activity occurs, and (2) where the harm is concentrated. These variations can be easily illustrated (see table 1).

A matrix based on the five locale-regarding variations in the table is not the only entrée into a legal-institutional analysis. Conflicts can usefully be grouped by reference to other variables, such as the num-

---

* Rights above the earth's equatorial belt are particularly valuable. In 1986 the equatorial nations proclaimed in the Bogota Declaration that their territories extended straight upward into the space orbits over their heads (comparable to the claim of the coastal states to an extension of control seaward), and that therefore any nation that wanted to occupy an earth-stationary spot above the equator would have to license it from the country below. The legitimacy of the claim has never been recognized. See chapter 9.

TABLE 1
Combinations of Act and Harm Sites

Site of Act	Site of Harm	Example
(1) Nation A	Nation A	A destroys a wholly internal critical habitat
(2) Nation A	Nation B	A's radioactive debris blows across the boundary into neighboring B
(3) Nation A	Commons	A's sewage invades the high seas
(4) Commons	Commons	A overfishes the high seas
(5) Commons	Nation B	A's tankers discharge wastes on the high seas, which invade B's beaches

ber of participants whose cooperation is required to bring the prob lem under control, and the generally perceived immediacy and intensity of the threat. But a quick overview of these five situation types provides a convenient introduction to the legal/diplomatic perspective, enabling us to identify, at least in contour, the starting point for concentrating appropriate policy responses. I will examine case 1 here; case 2, regarding transboundary pollution, is examined in chapter 3; the three situations involving the commons areas are gathered in chapter 4.

A Nation's Abuse of Its Own Environment:
Case 1, "Internal Affairs"

The paradigm for this first class of cases is the nation that overexploits its "own" nonmigratory living resources: its elephants or tigers or trees. Obviously, there are some borderline cases that test where the abuse of one's "own" resources ends and extraterritorial effects begin. Massive deforestation in Brazil has conspicuous implications for the global climate; forest cutting in Nepal increases the severity of floods in neighboring India and Bangladesh. In such cases the legal status may be viewed as shading over by degree into cases of the nation-to-nation and nation-to-commons types, which gives outsiders a stronger platform for objection. But assume here that any adverse physical effects beyond Nation A's borders are minimal. Have the outsiders any diplomatic or legal recourse?

The answer, under present international legal principles, is little. To support a lawsuit, or even to deliver a tenable diplomatic protest, the complaining nation generally has to prove some physical "intrusion" (or threat of intrusion) of its territory. A modest intrusion *may* qualify (especially if the implications of acknowledging it are modest): Mexico once got satisfaction on a diplomatic demand that the United States eliminate invasive odors from a Texas stockyard.[4] But absent some recognizable infringement of territorial sovereignty, many local factors that have in the aggregate considerable implications for neighboring states and the biosphere, such as population control, forest maintenance, agricultural techniques, public health administration, cultural attitudes (what constitutes a normal comfortable life?), are, as a practical matter, beyond the reach of international law.

One can of course lobby to change international law along lines that would more boldly check a nation's despoliation of its "purely internal" environment. There is some loose precedent in the developing body of International Human Rights law, under which nations are restricted in the treatment of their own people (and not merely of aliens and emissaries, jurisdiction over whom raised questions of international law from its inception). Certain aspects of the environment, such as rare species, could be considered part of the common heritage of humankind, wherever found, and their destruction, denominated as "ecocidal" acts, could be likened to genocide in the international community.[5]

But the prospects for such a movement are not good. For one thing, the history of International Human Rights enforcement has hardly been inspiring. And one would suppose it to be far easier to muster a worldwide consensus behind human rights and against genocide than to rally support for an anti-ecocide law. Cutting down trees does not arouse the same outrage as cutting down people. In addition, the tree cutters are at least offering economic development in return for their damage. For that reason even the environment-supporting 1972 Stockholm Declaration on the Environment recognized each nation's "sovereign right to exploit" its "own resources" pursuant to its "own environmental policies." At the Rio conference twenty years later, the developing world remained firmly behind that "right," which is understood in the context as a code for each nation's right to do with its environment whatever it wishes, without having to answer for internal effects internationally.

Short of a fairly radical shift in the underlying conception of sovereignty, there are only several options available to the outside world for responding to these "internal" situations.

Assistance in Planning

One approach is to help all nations recognize and implement conservation measures that are in their own best interests. That, in fact, is part of the task of many world organizations, such as the United Nations Environmental Programme (UNEP) and the Food and Agricultural Organization (FAO). As an illustration, spurred by world wildlife to discourage poaching, Tanzania has publicized the fact that an adult male lion in its wildlife parks generates $515,000 in tourist revenues over its lifetime; as meat and skin, it brings a poacher $1,150.[6] Other African states are now trying to arrange jobs at tourist resorts for populations from which poachers are most likely to be drawn.

But in many cases, a nation that has agreed to go through a set of planning discussions will emerge unpersuaded that its own interests and those of the outside world converge. In the particular situation at issue, preservation is frankly *not* the best internal policy. What, then?

Compensation for Conservation

One response is to pay the nation in question to adopt the desired course of action, either through direct compensation or debt relief.*

---

* The "payment" may take the form of each of a group of nations making a mutual concession. For example, the framework for the Convention on Wetlands of International Significance encourages any Nation A to offer Nation B, as a condition that B not destroy one of its wetland areas, an undertaking by A that it will set aside and preserve one of its wetlands.

39

The idea may at first sound odd or even affronting. But it is not unreasonable, and has been around for quite some time. Jomo Kenyatta, the founder of modern Kenya, is reported to have suggested at a 1961 conference in Tanganyika that if African wildlife was indeed a world possession, "the world could pay for it."[7]

The most widely discussed form of purchasing cooperation has been the debt-for-nature swap. The concept is almost invariably presented to the public wrapped in almost fairy-tale form. The story runs like this. A developing country, A, has borrowed funds by issuing debt. The face amount of the debt is $10 million. But because of the growing unlikelihood that A will be able to pay off its obligation, the debt is being deeply discounted in the international bond markets—to, say, $1 million. At this point an environmental organization offers to buy up the debt and to cancel it in return for some environmentally conscious action by A. At a bargain basement cost of $1 million, we are told, the environmental group is able to secure $10 million worth of environmental protection. It sounds too good to be true. And it is. If the debt was being resold for $1 million, then the value to A of the offer was at most $1 million.*

The true miracle surrounding debt-for-nature swaps does not involve any wizardry of financial leveraging, but a strange bedfellowing of lobbyists. A group of U.S. banks, pressed by bank regulators to unload drawers full of battered Third World debt, and a group of environmentalists combined forces to get a House of Representatives banking committee to write some obscure legislation that, with in-

---

* Indeed, what was being offered A was simply the value to A of not suffering in the future from the circulation of its deeply discounted debt. That amount could be measured in terms of eroded credit, or what A would have had to give up in a restructuring of its debt in 2005, etc. But the value would not, in any event, amount to $10 million because that's what the notes have printed on their face. A good test is to ask yourself which would A's government choose if it were offered: its bonds with $10 million stamped on their faces, but which are selling for $1 million, or $1 million in cash? The answer is $1 million in cash, for then the government would have the option: it could either go into the secondary market and buy up its bonds if it so chose, or it could use the cash for development. It could not do worse with the cash and it might do better.

triguing opportunities for currency exchanges and the inevitable convoluted tax-sweeteners that few are aware of and fewer still can follow, makes the whole thing run![8]

Far be it for me to complain: if everyone else gets special relief, why not Citicorp and Mother Nature, hand in twig? The idea of related swaps of various sorts is spreading, and taking many forms (cash-for-debt and equity-for-debt, as well as debt-for-nature). The Agency for International Development (AID) has put up $1 million to purchase Madagascar's debt in exchange for government support of local environmental groups. The World Wildlife Fund has arranged debt-nature swaps with the Philippines and Zambia. In a variant that could be seminal, Central American and Caribbean countries have been encouraged to take part in a debt-for-nature scheme called the Enterprise for the Americas Initiative. The Bush administration asked for authority to sell $5 billion in nonconcessional debts to foster investment in environmental or developmental projects. Each country would have to work out an environmental framework agreement with the United States, which would include the setup of an environmental protection fund and a local board for administering the fund.[9] Poland is trying to get the West to forgive 10 percent of its debt in exchange for an environmental protection program fund to tackle problems in the Polish environment.[10]

Whatever form the "purchase" of Nation A's cooperation takes, there are several problems with institutionalizing the arrangements. The most important challenge is to assure that the nation being paid will live up to its agreement. To illustrate, suppose that some sort of global environment fund (see chapter 9) agrees to provide Nation A $10 million not to disturb a critical habitat. The fund would not be in any legal sense purchasing a chunk of A's sovereignty; it would be contracting with A's government to keep the land in question undeveloped.

The problem is, what if A should accept the money or bonds and subsequently change its mind, claiming some intervening excuse from performance, such as a radical change in government? One answer is not to offer any lump sum up front, but to structure periodic payments so that the purchaser will hold the bond, but not cancel it until the issuer (the debtor nation) has done what it was supposed to do. Or, a simple rent agreement may do. Joan Martin Brown, an official with UNEP, has remarked: "The Third World says you're telling us not to do what you did to achieve your high standard of living. What are you going to do for us? Do you want to rent the trees from us? You know you can rent them for $1 billion a year in hard currency and we won't cut them down. . . . If you don't want to rent them, what is your quid pro quo? So far we [the advanced world] haven't been very forthcoming or very innovative."[11]

However they are worked out in detail, the pay-for-conservation arrangements are innovative—and possibly the most effective device for many of the "internal" situations.

Technology Transfer and Other
Environmental-Specific Assistance

To a developing country, a conservation payment may be construed as a bribe that it remain a Third World country, particularly if it leaves Nation *A*'s citizens with no more prospect, and no more self-esteem, than that of caretakers. In that case, another alternative is for the outsiders to provide technology that mitigates development's environmental blemishes. Such aid can be provided in anything from specific capital assets, such as water treatment plants, to training, such as in wildlife management, to the transfer of patents, royalty-free, for developmental technology that is environmentally benign (a prime request of the Third World).*

The "leapfrogging" of such technology to poor nations could prove to be one of the most effective strategies at our disposal. For donor

---

* For example, the developed countries are now making extraordinary advances in equipment for cleaner, as well as more efficient, uses of coal, e.g., fluidized coal beds. A nearly operational new technique, the PFBC (Pressurized Fluid Bed Combustion), promises to reduce SO_2 emissions by 90 percent and NO_x by 50 to 70 percent. *See* "Clean Coal Technology," *Science* 250 (1990): 1317. U.S. firms have recently announced a radically new light bulb, the E-lamp—aside from fluorescent, the first advance since Thomas Edison. It will be expensive to purchase but a bargain in energy savings—an astounding 75 percent—in the long run.

nations, such in-kind assistance has a political attraction over cash grants, that it is harder for the recipients to divert the aid into arms purchases or personal bank accounts. And even if it takes the form of patent transfers, it need not be done in a way that disregards inventor incentives; patent holders could be compensated, but look for their payment to a fund established by the developed world, rather than by the users in the LDCs. And, in any event, if thinkers like Amory Lovins are right, that what the LDCs most need transferred is technology "ordinary" in the West, such as decent motors, rather than esoteric, cutting-edge devices,[12] patents need not be a stumbling block.

Moreover, the "technology" to be transferred should be broadly construed to include legal and other institutional practices. Many nations can benefit from legal assistance in the drafting of antipollution laws, and from the training of local authorities in their enforcement. The point cannot be overemphasized: in many parts of the world, many environmental problems reflect not as much "natural" pressures as they do human folly inscribed in law. In Brazil, some of the worst excesses of deforestation, which achieved their heyday in the 1980s, have been curbed by reform in mindless land laws which, for example, had penalized claimants who did not put their mark on the land by burning it.[13]

Across the world, too, water is commonly in short supply not because there is not enough of it to go around, but because it is underpriced and distribution has been assigned on a use-it-or-lose it basis. As a result, farmers (and nations sharing an international waterway) fear they will only penalize themselves if they use water wisely.

This is not just a "foreign" problem. While it is fashionable and fun to blame California's water shortage on car washes and evaporation from swimming pools, Marc Reisner has pointed out that while in California today agriculture accounts for only 2.5 percent of the economy, it holds claim to 85 percent of the state's water. Cattlemen are California's biggest water users by far. Although the cattle industry grosses only $94 million a year, it swills 5.37 million acre feet annually to irrigate pasturage, the equivalent of 27 million of us maligned household users. Indeed, if the four largest water uses in the state—cattle, alfalfa (largely to feed the cattle), cotton, and rice—

were encouraged, by full cost pricing, to migrate to more suitable production sites,* and even if we imagined no crops were to replace them, California's agricultural gross would only decline from $14 billion to $12.3 billion, and there would be enough water for 70 million more people![14]

At home and abroad, there are many such sorry stories to be told. The good news is that they offer the prospect of enormous gains to be made outside of any transfer of capital, just from reforming perverse institutional practices, from water law to land use.[15]

The Conditioning of Economic Assistance on Environmental Performance

Another way to influence a nation's internal behavior is to condition general economic assistance in a manner that encourages borrowers, usually developing countries, to adopt policies that are environmentally sensitive. The idea is not to provide environmentally benign technology, such as fluidized coal beds, in kind, but to lend money—on conditions.

The principle influence has come from the various world and regional developmental banks—largely at the prodding of U.S. environmental groups. Increasingly, the pressure has led the banks to reject outright or demand a scaling back of proposed projects that had potentially catastrophic environmental implications. For example, the World Bank, spurred by the Environmental Defense Fund, reduced the scale of a huge agricultural scheme that threatened Indonesian rain forests.[16] The Asian-African and Inter-American Development Banks have been the target of similar pressure to cancel or condition loans upon the recipient undertaking environmentally responsible behavior, such as reforestation and soil erosion control. To strengthen this approach, lending institutions have been successfully

---

* Or better yet, if we just ate less beef (see Jeremy Rifkin, *Beyond Beef: The Rise and Fall of the Cattle Culture* [New York: Penguin Group, 1992] and fewer of our other fellow creatures while we were thinking about it, there would be a lot more protein to go around.

pressed to prepare Environmental Impact Statements (EISs) prior to approving developmental loans. If conscientiously prepared, the EISs should be a first step in identifying the conditions that ought to be attached.

Capital is not the only resource which it might be wise to provide only on condition. For example, there is a movement to assure that potentially toxic substances and technologies be sold internationally only if they are accompanied by instructions as to use and subject to conditions of use by the transferees.

Coercing Cooperation

These consensual conditions failing, there is a gamut of less temperate approaches, such as refusals to deal with—either to sell or to buy from—a country whose internal policies are environmentally unacceptable.[17] For example, the recently negotiated Basel Convention on the Control of Transboundary Movements of Hazardous Wastes and Their Disposal prohibits shipping such wastes, ostensibly for disposal, to any nation unprepared to handle them in an environmentally sound manner.

A harbinger of the refusal-to-buy approach is the Convention on International Trade in Endangered Species of Wild Flora and Fauna (CITES). Under CITES, the 103 covenanting parties agree not to sell or import animals, or the products of animals, that appear on the list of internationally endangered species. But boycotts need not be concerted. A single nation such as the United States, which represents a potentially major market for almost any good, can have a major influence through sensitive domestic legislation. Recent U.S. threats to invoke trade sanctions against Japan have led to a Japanese commitment to eliminate importation of Hawksbill turtle shells.[18]

Boycotts, however, may not only be too abrasive, they may also prove too blunt. A threat to hold hundreds of millions of dollars in trade hostage will be credible, if ever, only if the environmental misdeed is blatant and egregious. In ordinary circumstances, milder, more balanced measures may be preferable. One bill brought before the U.S. Congress would simply have suspended the duty-free import of wood articles (available under the so-called General System of Preferences) to any hardwood-exporting country that did not implement an appropriate reforestation program.[19] If demand cannot be eliminated by such product pressures, it can at least be reduced.

All trade sanctions, however, from boycott to pressure tactic, raise several problems that have to be borne in mind.

First, their legal status is cloudy. Recently, the United States banned the import of tuna caught by Mexican fleets that destroyed more dolphins than allowed under the United States' Marine Mammal Protection Act. Mexico responded by filing a complaint under the General Agreement on Tariffs and Trade (GATT), maintaining that the U.S. ban was a discriminatory trade practice. A GATT panel upheld Mexico.[20] The ultimate significance of the ruling is not presently clear,* but it at least suggests a setback to the environmentalists, that any nation's unilateral attempt to make its environmental policies felt extrajurisdictionally via trade sanctions may come into conflict with trade agreements.

Even if the legal complications can be finessed,** there is an economic hitch. Because the countermeasures make fewer of the targeted products available on world markets, the price of the contraband is driven up. The effect may be perverse. If rhinoceros horns are banned from trade, each horn becomes all the more valuable, thereby making poaching more lucrative. The objection is not fatal—else we would not ban the sale of anything. Indeed, the higher prices may also foster development of substitutes. For example, restrictions

---

* Mexico, sensitive to bad environmental publicity as the North American Free Trade Agreement approaches final confirmation, has agreed not to press its victory in this small battle so as to allow a far-reaching compromise to be worked out under the auspices of the IATTC (Inter-American Tropical Tuna Convention).

** The GATT panel that held for Mexico intimated that what made the U.S. action offensive was the unilateral character of its efforts to project its own law abroad, and that authoritative collective action, such as a multinational boycott under the framework of CITES, would be acceptable. All agree that the U.S. can, without violating GATT, restrict imports of agricultural products such as fly-infested fruit, which would adversely affect the domestic environment; the dolphin issue involved imports that were produced by a practice offensive to domestic sensibilities, not health or safety.

on the importation of monkeys, once commonly (and carelessly) used in medical laboratories, has had the unintended effect of driving prices up, with the result that more researchers are shifting to other models, even nonanimal ones.

Another point to remember is that many pressure strategies are distinctly situational. Heavy importers of endangered species products, such as Hong Kong and Japan, are potentially the most effective sanctioners—but may also be, for the same underlying reasons, the least inclined to engage in boycotts and sanctions. Similarly, if the refusal to supply capital to environmental wrongdoers is to be adopted as a strategy, one must remember that while it may provide leverage against Third World borrowers like Brazil, it will be of no avail against wealthy nonborrowers, such as Japan and the United States. Ousting an offending nation's fleet from coastal waters could be an effective sanction for coastal states to mete out, but is unavailable for the landlocked nations.

And, finally, there are very few circumstances likely to cause the major players in the world community to apply strong sanctions against a nation, however poor, on account of some "internal" delinquency. Deforestation is not aggression. We are inclined to be disappointed by environmental despoliation. But the country that reduces its biological diversity is less likely to be branded as an international culprit than to be thought of as another poor country just trying to grow by doing what it has a traditional right to do. Too much pressure is likely to make the target feel more sinned against than sinning, straining international relations unnecessarily.

Realistically, the most muscle-flexing strategies are better reserved for situations where the actor is seriously affecting other nations, or violating the global commons.

Unrestricted Economic Aid as an Environmental Policy

For the wealthy resource-providing nations, environment-specific assistance and conditioned financial relief offer clear advantages: by providing aid in kind, or attaching strings to their assistance, aid providers add assurance that the recipients will respond to the providers' liking. For that very reason, the availability of tightly controlled assistance is calculated to make the wealthier countries more openhanded about providing the assistance. But from the recipients' perspective, strings-attached aid is seen as a way of imposing the developed countries' agenda on theirs. The interest in poor countries may well include the global environment. But for most of the LDCs, the specter of a theoretical and remote environmental collapse is in general less foreboding than the stagnation and poverty that they experience every day. Indeed, as we have seen, while the wealthy nations have the luxury of worrying about skin cancer and global warming in fifty years, the poor are still trying to eliminate malaria and dysentery, to get themselves fed, and then to proceed, hopefully, with building a basic economic infrastructure, including housing, communications, and roads.

How ought the developed nations respond to these desperate demands? The first possible answer is: by meeting them on their own terms, as gambits in hard bargaining. We need their cooperation for what we value; they need ours for what they value. On this sheerly hard-nosed view, their cooperation is to be purchased for whatever it is worth to us, in getting them to clamp down on terrorists or whatever.

The second answer is: with sympathy. We have to recognize that truly impoverished nations may not yet be at a stage where they can "afford" to put their environmental consciousness into practice. How environmentally conscious were the United States and Western Europe in their industrializing phase? On this view, increased aid ought to be made out of compassion, with a frank realization that even if the environment suffers, the alleviation of human suffering comes first.

The third response is to increase outright economic aid in proportion to some hope that, at least in the long term, environmental protection to mutual advantage will follow. A recent report by the World Bank argues impressively that just as the cure for poverty is economic growth, so too is economic growth both part and parcel of, and a staging platform for, repair of the environment.[21] Speaking in traditional economic terms, the Bank points out that many of the

48

world's poor simply cannot afford to ease up on the environment, even if to do so would be, by Western standards, an obviously worthwhile investment. For example, among poor farmers in India the Bank found an implicit discount rate of 30 to 40 percent, that is, the farmers were so pressed they were unwilling to make an investment in soil conservation or tree planting unless it would treble its value in three years.[22] With more resources, and more goodwill toward a community of nations, they will be better able and more willing to relieve pressure on the environment—their own and the globe's.

Increased levels of general economic assistance now, with no strings attached, may simply be a worthwhile price to achieve environmental stability and world harmony in the long run.

Transboundary Pollution

IN THE SITUATIONS just reviewed, the brunt of a nation's actions fell on its own internal environment. In this chapter we proceed to consider the second class of cases: those that arise when a nation engages in an activity within its own territory which casts its influence—in the form, for example, of toxic gases or radioactive debris—across the border into a neighboring state.

Of course, a nation's air or water pollution will ordinarily work some damage within its own territory before it crosses the frontier into a neighboring state or the commons area. This self-inflicted portion of the damage gives each polluting nation some incentive to install filters or scrubbers to clean up its own act. But even if the polluter, A, has the wherewithal and will to control the damage, its self-interests motivate it to abate the fluxes or reduce the risk only to the extent that the marginal costs of abatement do not exceed the marginal benefits *to it*. In other words, any nation will be inclined to initiate unilateral domestic action that may incidentally reduce extra-territorial damage but only to the extent that national self-interest requires, and not necessarily up to the mutually (or globally) desirable levels. Each nation's preferred alternative, when it has the option, has too often been the construction of taller and taller smokestacks, so as to carry its fumes out across someone else's turf.

In this case, A, the transgressor,[1] can less convincingly dismiss the outside world's criticisms as a meddling in "strictly internal affairs." While the threshold standard for significant external effects is not always obvious, or rational,[*] in general it is clear that the stronger the

---

[*] A nation that engages in massive deforestation, impairing the earth's capacity to draw CO_2 from out of the atmosphere, might be doing more harm than another nation that positively launched a small amount of gas or debris outward. But because of conceptions of trespass and nuisance that are embedded in the law, it is unlikely that a nation that reduces what it draws *inward* across its frontiers faces the legal liability of the nation that casts something *outward*.

external effect, the less persuasive are the polluter's invocations of sovereign prerogative.

Accordingly, there are any number of international declarations that agree in essence that

> (1) A state is obligated to take such measures as may be necessary, to the extent practicable under the circumstances, to ensure that activities within its jurisdiction or control
>> (a) conform to generally accepted international rules and standards for the prevention, reduction and control of injury to the environment of another state . . .; and
>> (b) are conducted so as not to cause significant injury to the environment of another state.[2]

In line with such declarations, and the sources from which they are assembled, there is some basis to argue (albeit with more support among academic commentators than in any case decisions) that, depending on the circumstances, A, in planning its activities, is or ought to be obliged to take B's interests into account, or even consult with C, or, in the case of imminent perils, such as an escaping toxic cloud, give notice to those who may be imperiled. Most important, formal diplomatic complaints or even lawsuits, unavailable or unavailing in the "internal" cases we examined earlier, are at least *theoretically* available in some of the transboundary circumstances we are looking at here. Even so, the practical significance of all these measures—absent material advances in the law—is apt to fall short of stemming the flow of transboundary pollution.

To illustrate, imagine that A is in the planning stage of siting a nuclear power plant near its border with B. Let us review the obligations and remedies that *might* come into play, either at present or with relatively modest modifications that negotiators could conceivably work into the framework of existing law.[*]

---

[*] We are dealing in the text with situations in which the hazarded nation cannot effectively block the perils through its own laws and institutions. Sometimes international law permits a country to adopt self-protective measures. For example, B can condition the entry of ships into its coastal waters on their being constructed according to B's domestic safety standards. Even in the case of more typical transboundary pollution, each nation's domestic courts have some power to entertain

THE PREVENTIVE TECHNIQUES

A Domestic Accounting of the Impact
of A's Actions on B's Interests

A, in the course of planning the project, could be expected to assess and, while under no legal compulsion, account in a "good neighborly way" for the effects on B. As for the United States' own position, in the past, both the National Environmental Policy Act (NEPA) and the Endangered Species Act (ESA) have been understood to require that agencies planning actions that can be expected significantly to affect another state's environment have to prepare an EIS, just as though they constituted major projects in our own interior.* As a result of a lawsuit brought by four U.S. environmental groups, the Agency for International Development—the principal channel of bilateral development assistance from the United States—has adopted regulations of slightly less demanding depth.[3]

What an EIS requires varies under federal and state laws, but in general the procedure applies to any major action, such as the building of a dam or cutting of a roadway through a forest, that may have some material adverse effect on the environment. The EIS procedure does not empower anyone to tell the proponents of a project that they can or cannot go ahead with their plans or even what modifications are required for approval if environmental threats show up. Instead, the idea is to force the proposers to identify and assess the potential impact of the proposed action on the environment, and compare that impact with the environmental impact of alternative undertakings that might meet the same objectives. The report is then circulated to appropriate agencies for comments, and if controversy is considerable, may be made the subject of public hearings. Of

---

suits against foreign polluters, particularly if the foreign culprit is a private firm rather than an agency of the state (because the state and its offiers may be able to take refuge behind sovereign immunity or the act of state doctrine).

* Whether the United States will continue to set a good example in this area is currently an issue before the Congress; see discussion below.

course, requiring consideration of environmental impacts does not guarantee full environmental protection. The hope is to encourage reflection before the work proceeds. That is a time when minds are most flexible, and the costs of making modifications the lowest.

The process has its critics. It entails time and money, which the public ultimately bears in higher prices for housing or fuel or whatever. And when all has been said and done, there is not much gained when the EIS preparer has gone into the task with its mind made up, in bad faith, and with a squadron of good lawyers. Even so, the political judgment remains that the costs are worth the heightened environmental sensitivity that they buy. Recently, the U.N. Economic Commission for Europe (ECE)[4] endorsed impact assessments for fifteen activities, and other countries have either adopted or are actively considering comparable procedures.[5]

Unfortunately, domestic experience provides little hard evidence on which to judge the effectiveness of the EIS technique applied to extraterritorial activity. A few suits were brought in the 1970s challenging extraterritorial action on the grounds that U.S. agencies, acting abroad, had failed to abide by the EIS provisions. One case sought to prevent the sale of a nuclear reactor and related materials without an examination of the environmental safeguards in the recipient nation. It failed.[6] Another action was brought by a group favoring the decriminalization of marijuana, challenging United States participation in herbicide spraying of marijuana and poppy plants in Mexico without preparation of an EIS. The District Court agreed that the agencies involved were in violation of the NEPA for failing to prepare, circulate, and consider a detailed environmental impact statement on the effects; but the court refused to enjoin the government from proceeding until an EIS had been prepared.[7]

Although this history is unpromising—certainly so, if our only measure is courtroom victories—one cannot conclude that EIS-type requirements are utterly ineffective. One has to remember that the cases actually litigated give us only a one-sided glimpse of the true impact. What we do not find in the court records are the cases that never got to court—the cases where the agency, faced with the enforced soul-searching of an EIS procedure, withdrew or modified its

53

original plans because it could not prepare an impact statement that it could live with—or that it could defend against the more vigilant environmental groups without an extended court fight. That, it should be kept in mind, is the true intention of the EIS.

In all events, the Bush administration came out opposed to a Senate bill that would clarify and secure on a legislative basis the obligation to continue the extraterritorial application of NEPA, and the Supreme Court, in its treatment of the reach of the ESA, has made the situation cloudier—and the stakes even higher yet. As for NEPA, the dispute has been whether the procedure is mandated by Congress under the legislation itself, or whether it is only contingent and stems from an executive order of the president.[8] If the latter, as the Bush administration is maintaining, the continuance of the practice depends on nothing more than presidential prerogative and may be withdrawn by the executive. A bill to affirm that the requirement is a legislative demand is presently in Congress.[9] While that bill was proceeding, in June 1992, the U.S. Supreme Court, in *Lujon v. Defenders of Wildlife*—a case aimed at testing the application of the ESA to activities in foreign states—stunned the environmental movement not only by casting doubt on whether the ESA did so apply, but also by refusing formally to reach the question on the grounds that the environmental group making the claim lacked legal standing to bring it before the Court. In Justice Anthony Scalia's far-reaching view, Congress might not even have the constitutional authority to empower citizens' groups to raise questions about the integrity of foreign environments when they cannot show true injury to their own interests, as opposed to those of a distant environment. Thus, in regard to the extraterritorial scope of the ESA, there are now two unresolved issues, the one—casting its shadow over the power of environmental groups to institute litigation—even more portentous than the other. Both controversies deserve greater attention than they have received. If the United States weakens its own requirements, it is hard to believe that other nations will fashion—or, at any rate, follow—comparable requirements of their own. The result will be a decreasing sensitivity to the environment beyond each nation's borders, just when we ought to be moving in the opposite direction.

Notification

Whether or not Nation A is obliged to go through an internal EIS-type procedure, we could require A to notify B of its plans in advance. There is some support for the view that such an obligation already exists in international law, at least in the context of a nation altering the flow of an international waterway.[10] With respect to air pollution, the Organization for Economic Co-operations and Development (OECD)[11] has adopted the "principle" that any nation planning a project that can be expected to have major transboundary impact notify and consult with those who may be affected.[12]

In most situations of potential transboundary conflict, one might suppose that notification, in itself, would be a rather hollow gesture. A plan to alter the flow of a river or to site a power plant—in effect, any project of a sort that might be expected to cause "significant harm" and therefore to trigger the obligation—is something neighboring states can probably discover by reading their neighbor's newspapers. But when international lawyers speak of notification they mean something more than bare notice; in the diplomatic context, notification is calculated to initiate discussions and public spotlight—both of which may cause the nation proposing the action to modify its plans.

Notification has its highest potential where it requires a disseminated warning of imminent perils. Such a duty-to-warn is now accepted in at least one area. After the accident at the Chernobyl nuclear power plant, the Soviets acknowledged a "moral responsibility" for the transboundary pollution but refused to discuss legal liability. However, largely as a result of the Chernobyl experience, the membership of the International Atomic Energy Agency (IAEA), including the USSR, mutually agreed to give notice of radioactive emission likely to affect other nations.[13]

The adoption of the nuclear agreement should inspire broader applications of the same principle. During the drafting stage of the Stockholm Conference in 1972, there was a proposal under consideration that would have entreated notification in all sorts of developing transboundary perils, including what happened at Chernobyl.[14]

Unfortunately, the wording was watered down just before the final vote. The way things stand, the next time there is a toxic cloud of the sort released in Bhopal, India, in 1984, there will be no clear obligation in international law to warn neighboring states to take reactive measures, such as the evacuation of populations, should the cloud head toward an international frontier.

The time has certainly come to press for a formal adoption of a provision in the spirit of the language rejected in 1972. Such an agreement is probably well within reach. The original rejection was based on some confused interpretations, and since that time the potential backing for such a provision has clearly grown.[15] Of course, in the circumstances of imminent danger, an obligation of advance notification can do little more than to soften damage that is already in the wind. The ideal is to prevent the tragedy in the first place.

Consultation

Some authorities have suggested a slightly stronger obligation than a duty to inform: that in some circumstances Nation A might have an obligation to sit down with B and discuss projects posing considerable risk of transboundary harm. Some commentators have called such a principle "an evolving rule" of general international law, applicable to all countries regardless of their consent, but that is certainly a wishful overstatement.[16] On the other hand, formal consultation agreements have been expressly included as part of particular compacts.[17] As mentioned, the OECD has promulgated a nonbinding consultation principle for projects that can be expected to affect transboundary air quality. And the United States has long-standing consultative arrangements both with Canada, as part of the International Joint Commission (IJC), and with Mexico, as a link between the United States' EPA and Mexico's FEDESOL (the acronym for its Ministry for Social Development).

Advance consultation, on a formal basis, is a process that should be more widely fostered. In 1986 several European countries advanced hope for a treaty requiring nations planning to build nuclear plants to consult with their neighbors on safety standards.[18] But nothing thus far has come of it, lamentably. Indeed, it hardly seems excessive to insist that a country planning to build any hazardous facility—nuclear or otherwise—near an international border should consult first with its neighbor as a matter of course.[19] Such consultations may or may not halt the plant siting. But merely forcing dialogue may result in the construction of a safer plant than otherwise, or in the plant going forward but with another irritant to the complaining state, such as river pollution, being mitigated as a quid pro quo.

Veto Power

Veto power is the highest level of accounting that the law could require: to force the nation proposing hazardous action to secure the prior approval of the hazarded states. A few—a very few—narrowly drawn bilateral and regional treaties have adopted such a requirement.[20] But nothing of the sort is claimed to exist under general principles, and there appears to be little enthusiasm for veto control in any areas at this time.

THE REACTIVE TECHNIQUES

The four techniques just reviewed all aim to divert the hazard-causing nation from a projected course of action before harm has been felt by its neighbors. To a traditionally trained domestic lawyer, they are all "soft" techniques, in the sense that they are unbacked by forceful remedies for noncompliance. Even if a state were "obliged" to consult, one cannot imagine it being held accountable in damages, or its reactor being deactivated, for a failure to comply.*

But suppose now that the harm has occurred: Nation A did notify B of its intended plans, or even sat down with B and discussed them in advance; nonetheless, A went ahead and B is injured. What recourse has B, then?

---

* One should keep in mind, however, that a nation which wrongfully failed to consult might face diplomatic countermeasures, such as a decline of reciprocal obligations toward it, and some erosion of face in the world community. These are traditional "remedies" in international relations, and not to be lightly dismissed.

Diplomacy

As a starter, there is the possibility of diplomatic maneuvering. To the general public, the passing of diplomatic notes sounds vain, starchy, and vaguely Austro-Hungarian, but a lot of international business still gets done the old-fashioned way. In 1961 Mexico delivered to the United States a note protesting odors from a Texas stockyard, a relatively minor form of nuisance that probably could not have been made the object of an international lawsuit; nonetheless, the United States acceded, apologized, and remedied the situation.[21]

In assessing the region of transboundary conduct over which such diplomatic activity may be effective, one must point out that in the U.S.-Mexico stockyard case the source of the trouble was politically and economically easy to correct. The doors of no major plant, much less of an industry, are readily to be closed by a *note*—as our acid-rained-upon Canadian friends know all too well.

But on the other hand, "diplomacy" offers the opportunity for a wide variety of imaginative solutions. To illustrate, for years before reunification, West Germany had been suffering from pollution invading from less prosperous East Germany. Shortly before reunification, and as result of diplomatic maneuvering, the East Germans agreed to clean up a number of sites in the East, including the notorious Buna chemical complex, with the West Germans agreeing to pick up the lion's share of the projects' costs.[22] The payments may have been viewed as simple aid, but it is also possible the West Germans calculated that the spillover from East German pollution was so bad that a Deutschemark spent in the East bought even those in the West more cleanliness than the same Deutschemark spent on marginal cleanup improvements at home.

Litigation

In any particular case, if diplomacy fails, two cases (as well as numerous international principles) suggest the possibility of litigation, of *B* taking *A* to court: *Trail Smelter* (*U.S. v. Canada*)[23] and *Corfu Channel* (*United Kingdom v. Albania*).[24] The former grew out of a protracted dispute between Canada and the United States in the 1930s over fumes from a Canadian lead smelter that were wafting into the state of Washington. The United States initiated an international arbitration and won, the tribunal declaring that "under the principles of international law . . . no State has the right to use or permit the use of its territory in such a manner as to cause injury to the properties or persons [of another State], when the case is of serious consequences and the injury is established by clear and convincing proof."

In the *Corfu Channel* case, the International Court of Justice (ICJ)* awarded a judgment to Britain from Albania after two British warships, making innocent passage, were sunk by Albanian mines, the ICJ enunciating the "obligation of every state not to allow its territory to be used for acts contrary to the rights of other states."[25]

The language from these two cases, however general, has been read by many international lawyers as providing a strong platform for transboundary litigation. I am less sanguine about litigation, however, particularly about cases instituted under customary principles of international law (that is, international law unqualified by specially tailored treaties and conventions that require the consent of all signatories and are discussed in chapter 5). Indeed, if the basic law provided a sufficient foundation for a nation with pollution complaints to go into court and get relief right now, why are more nations doing so? It is nearly half a century since *Trail Smelter* was decided. Surely the volume of transfrontier pollution has not diminished. Indeed, if the existing principles are so favorable as some of my colleagues believe, they need to explain why the Canadians don't turn *Trail Smelter* around and sue the United States for the fluxes that result in acid rain.

Part of the answer has to do with the attitude of public leaders, who are the ones to decide whether to initiate nation-to-nation litigation or pursue ordinary diplomacy. The short of it is, national leaders

---

* The ICJ's fifteen judges are elected for nine-year terms by majority votes of the Security Council and the U.N. General Assembly. Under the Statute of the Court, they are to represent the "main forms of civilization and of the principal legal systems of the world," a goal that is advanced by assuring a broad geographical representation.

are much less favorably disposed to "go to law" than are lawyers and law professors. From the statesman's point of view, to go to law is to transfer control to another profession, one that is hard to predict (much less to control!) and which threatens to make a specific, multifarious situation turn upon some general hoary principle—in their worst fears, incanted in Latin.*

Moreover, each nation knows that even if it is successful, it will be fortifying doctrine and practice that may come back to haunt it. The plaintiff may expose itself to countersuits regarding not only its own pollution, but other international grievances, present and future. This would feed a trend that most international lawyers favor, and that the idealist regards as vital: a growing ambit of general universal principles that all states recognize. But that is a development which those whose freedom of maneuver is most threatened—the political leaders—are likely to approach with the most caution. Indeed, viewed not merely as a matter of self-interestedly protecting one's "turf," but as a matter of protecting the environment, there are many circumstances in which diplomatic maneuvering may provide the best solution.

In fact, despite the plaintiff United States' success in *Trail Smelter*, a quick examination of the circumstances of that case provides a useful lesson in why, in more typical situations, a litigation strategy is almost certainly destined to be inadequate.

The Absence of Compulsory Jurisdiction in the World Order

First, *Trail Smelter* was an arbitral proceeding to which both nations consented. Nations do not have to respond to a summons and complaint, the way individuals do. In the international community, for a

---

* From the statesperson's perspective, the prospect of international arbitration (such as in *Trail Smelter*) may not be quite as unsettling as going before the United Nations International Court of Justice (ICJ). The ICJ hands down edicts based on whatever general principles of international law the justices find to be applicable. In arbitration, by contrast, each nation can select the arbitrators and agree in advance to the rules by which the arbitrators are to be guided.

nation to allow other nations or their citizens to sue it is a matter of choice: the sovereign cannot be sued unless it consents. Depending upon the conditions of its acceptance, if any, of ICJ jurisdiction, the damage-causing state may simply refuse to appear.* In *Trail Smelter*, the controversy was no more than a minor sticking point between the nations. Canada wound up being assessed a trifling $78,000 in damages. Other nations, in the face of more consequential disputes, might not be so quick to consent. In 1973 New Zealand and Australia called upon the ICJ to declare that French nuclear-weapon testing in the Pacific, which hazarded downwind populations, constituted a violation of international law. When, despite France's refusal to appear, the Court proceeded to grant interim relief, France announced that it planned to proceed with the round of tests anyway, and, in pique, withdrew its consent to the Court's jurisdiction in the future.[26]

Unclear Standards of State Accountability

Suppose that a nation which is sued does consent to jurisdiction. The next question is: For what actions is it accountable? It is widely accepted in international law that states are responsible for compliance, by themselves and by persons under their jurisdiction, with norms of international law, including those of environmental conduct. There remain, however, largely unresolved questions regarding the standard of conduct by which a defendant state is to be judged liable. Is a nation to be charged for all pollution that drifts across its neighbor's lands, however innocently and diligently it behaved (that is, in legal jargon, to be subject to strict or absolute liability)?

Here one does well to refer back to the qualification that the United States impresses upon state standards in the excerpt set forth at the beginning of this chapter: "A state is obligated to take said

---

* Note that regional courts may have more power in these regards than the ICJ. The European Court can oblige—and perhaps enforce—an environmental directive against members of the EC.

measures as may be necessary, *to the extent practicable under the circumstances*" (emphasis added.) This seems to imply that if measures to reduce acid rain are too expensive (too expensive to whom, and how weighed?), there is no obligation to mitigate. Another alternative is that *A* be liable only for damages that arose because it did something positively wrongful, such as breaching "a generally accepted international rule or standard," for example failing to give *B* timely notice of a spreading radiation peril as required by the new (post-Chernobyl) agreement under the auspices of the IAEA. Again, in *Trail Smelter*, the issue simply was not pressed: for that particular controversy, Canada agreeably stipulated that the arbitrators should apply, and it was prepared to be bound by, the standards of U.S. law.

Currently, the influential International Law Commission is examining the issue of state responsibility, and its drafts indicate some inclination toward strict liability.[27] That would be the most favorable standard for the plaintiff, since it dispenses with the requirement of proving the defendant's fault. On the face of it, strict liability would strengthen the position of nations injured by transboundary pollution. But writing strict liability into the law could backfire against those injured. A nation whose pollution has been challenged but which is denied, under the strict liability doctrine, an opportunity to defend on the grounds that its actions were reasonable or that the damage was unforeseeable, will be that much less likely to consent to be sued.*

Indeed, at the present stage of world community, the principle of strict liability for all transboundary damage is so impractical—because transboundary pollution is so widespread—that the effects of adopting such a principle would almost certainly have to be offset by raising the threshold of what constitutes a legally actionable

---

* The part of the "strict liability" movement that aims to put pressure on multinational corporations, not states, has more promise; corporations, unlike nation-states, cannot counter by invoking sovereign immunity. See U.N. International Law Commission Split on Transboundary Pollution Liability, *International Environment Reporter (BNA)* 11 (1988): 166.

level of "damage," or by putting the complainant under a stronger burden to prove that the defendant's activities were the cause of its injuries.

Problems of Proof

Third, the questions of proof are often too daunting to overcome. In *Trail Smelter*, the United States had caught Canada with a smoking smokestack. But more typically, a plaintiff nation will be put to a controversial proof: was *B*'s territory invaded by some agent, call it chemical *z*, and if so, *which* nation's *z* is to blame? (*B*'s own?) Is *z* by itself benign, and mischievous only in combination with another agent, *y*, that had been unloosed by someone else? Even should the courts adopt a standard of strict liability (liability without proof of the acting state's negligence), the plaintiff would still have to prove the causal connection between the activity complained of and the injury received.

Inadequacy of Relief

Even if the defendant appears and is held accountable, injunctive relief, a principal prop in enforcing domestic environmental law, is rarely available in international forums.[28] Damages, when awarded at all, tend to be calculated more miserly than in comparable cases under prevailing U.S. practice. Damages to the environment that do not show up as property losses—for example, the loss of a wetlands area that has no commercial value—are certainly apt to go untallied. Punitive damages are unavailable, and some portion of what U.S. courts would regard as fairly commonplace elements of ordinary damages, such as the reduced value to pollution-blighted businesses and property, may be rejected, as they were in the *Trail Smelter* case, as "too remote and indirect to become the basis . . . for an award."[29] When one considers the costs and delays of suit, and the likelihood that most plaintiffs will ultimately lose anyway, the level of legal damages a transfrontier polluter realistically faces falls far short of forcing it to face up to, and therefore efficiently adjust for, the full measure of the costs its polluting activities are imposing on its neighbors.

Defects in Enforcement

Litigation is notoriously slow. Even though Canada did not contest its responsibility in *Trail Smelter*, the proceedings lasted over twenty years. Irreversible damage may proceed far faster than the legal system that is on its trail. Indeed, even if the courts finally hand a damage judgment down, there is no assurance it will be enforced. The rarely mentioned punch line to the *Corfu Channel* case is that, forty years later, Great Britain has yet to see a penny of its £700,000 judgment.[30]

Suboptimal Settlements

One final warning about shortcomings of bilateral litigation. Many lawsuits, perhaps most of them, are settled out of court. But there is no assurance that the level of pollution mutually agreeable to two disputing nations—what A and B are prepared to settle for—will be the ideal level from the perspective of other nations, not party to the litigation, who may suffer the pollution either directly, or indirectly through atmospheric or oceanic degradation. One nation's pollution complaint is likely to be settled by reference to its neighbor's own pollution complaint—plus a few items of contention the diplomats regard as more pressing: trade, currency, and military assistance.

What this last point underscores is the importance, once more, of public opinion in emphasizing environmental priorities and then rewarding governments for delivering on them. Even if the pressure falls short of producing immediate radical reforms in formal international law, it can have impact in shaping what are called "state practices"—which in turn redound to influence the norms and expectations that govern international behavior.

Are Such Suits Bootless?

None of this is to say that international pollution suits ought not to be pressed in the transboundary situations, even in the face of the difficulties. Such suits can have a salutary influence both on the actors and, ironically, on the law, even if the "defendant" state refuses to appear. Such was the apparent result of the 1974 *Nuclear Tests* cases (*Australia v. France, New Zealand v. France*). France (which, like China, has refused to sign the 1963 Nuclear Test Ban Treaty) initiated in 1966 a series of nuclear weapons tests from French territorial bases in the South Pacific. These led to protests from Australia and New Zealand, both of which were downwind of the test range. France rejected the two nations' diplomatic protests on the grounds that atmospheric tests, of themselves, were not unlawful, and that any radioactive depositions they had experienced did not cross a legally significant threshold. Thus rebuffed, Australia and New Zealand filed suits before the International Court of Justice. France, maintaining that the ICJ was "manifestly not competent to try the case,"[31] took no further steps in the Court—but made it clear it intended to continue testing. At this point, the plaintiff nations persuaded the ICJ to hear them out even without France's participation, and to enter some form of interim measures of protection, ostensibly while the formal authority of the Court to proceed could be resolved.

In 1973 the ICJ, by a slender 8–6 vote, took the opportunity to urge the French "to avoid nuclear tests causing the deposit of radioactive fall-out on Australian territory."[32] France disregarded the Court, pursuing its test schedule to conclusion. The French government then announced that it intended to cease all further atmospheric testing by the end of 1974. The Court thereupon ruled that no further proceedings were required, the Court construing the French government's announcement of test cessation as a binding commitment to cease disturbing the atmosphere, and therefore as a sort of concession to the demands being made upon it.

Thus, while no definitive judgment was ever reached in the *Nuclear Tests* cases, there were positive results. To begin with, the ICJ managed to hand down an opinion that strengthened the position of environmental plaintiffs in the future.

Moreover, any such ruling as in the *Nuclear Tests* cases, even if it does not command immediate obedience by one of the parties, is a way of legitimating and broadcasting attitudes that can become important factors in shaping world relations, even if they do not establish formed precedents. The more credibly a doctrine raises the specter of liability should the parties ever get to court, the stronger the bargaining hand of a complaining nation in out-of-court diplomatic negotiations.

CONSENSUAL TRANSBOUNDARY POLLUTION

Up to this point, I have focused on situations in which B has been an involuntary victim of A's pollution. But the growing international commerce in toxic wastes is raising questions with another twist. The industrial countries, which produce the bulk of the hazardous waste, are running out of deposit sites. The LDCs, which generate less waste of their own, are hard up for cash and are often willing, for a price, to make some of their uninhabited spaces available for deposit. In these circumstances, the state accepting the waste, as distinct from the victim in the involuntary pollution case, is consenting willingly.

Philosophically, the issue raised is within the realm of paternalism: when, and with what justification, should a society interpose itself between a willing seller and a willing buyer? Law school professors still play with the old prize-fighting cases: If two fighters are willing, for a fee, to accept the risks of injury, why should the state ban the fight? The question is not rhetorical; but neither is the answer clear-cut.

There is at least one facet of the dilemma that virtually everyone would agree to: the law should at the least assure that the buyer is adequately informed of the risks. And that is a position inscribed in the Basel Convention on the Control of Transboundary Movements of Hazardous Wastes and Their Disposal, which went into force in 1992.[33] Signatories, including the United States, have agreed not to ship such wastes unless the recipient state, as well as states through which the wastes must transit, are notified in writing about the shipment and given opportunity to consent or refuse. Without such notification and consent, the shipment may not proceed.

The next step is somewhat harder. Suppose that the buyer nation has been fully informed of the risks but is willing to proceed for the price. The Convention continues to prohibit the transaction if the shipping nation "has reason to believe that the wastes in question will not be managed in an environmentally sound manner"—even though, ordinarily, only the consenting storing nation would suffer.* Although the storage in those circumstances would be, strictly speaking, voluntary, it seems to me that the Convention is clearly right. Much of the Third World notoriously lacks both regulation and expertise in hazardous waste disposal. If the shipper has that expertise, and knows that there is going to be a problem, then its pursuance of the prospect can be viewed as making it an accomplice, morally speaking, in an environmental tort or crime.

The question gets toughest at the next level, where the receiving nation is fully informed and willing and the shipper has no reason to believe that the handling will be unsafe. In those circumstances, the Convention allows the sale to go forward—and that is the crux of the lingering and poignant controversy.

On the one hand, there are those who point out that waste containment is inherently risky and believe that the practice of hiring someone else to store your risky garbage should be banned, period. Their position, essentially, is the predominant position of Anglo-American law in regard to, say, prostitution or, perhaps more in point, consent to a battery: the law refuses to recognize a contract in which a poor person agrees to be beaten up for a price.

My own misgivings about the consensual transboundary shipments grow from a slightly different consideration. To me, risky contracts are not the problem per se; otherwise coal mining and tunnel

---

* The analysis varies if there is a risk that wastes stored in the storer's territory were likely to leak via a waterway or underground aquifer into the territory of a neighboring nation not party to the storage agreement between the shipping nation and the storing nation.

digging and a lot of other activities would have stopped long ago. The toxic storage cases present another layer of complication. The prostitute and boxer—and miner and tunnel digger—are deciding to take risks for themselves. But here we have considerations of agency: the leadership of a Third World country is deciding to take risks that will fall on others, its citizens, present and future. If we assume that we are dealing with a representative government that works perfectly smoothly, then, of course, the cases are the same. But that is quite a heroic—and blind—assumption. Representative democracies have not done so well on waste storage themselves and few of the countries involved on the storage end are thoroughgoing democracies. Realistically, warnings about the shipment's hazards will often be restricted to the recipient country's leadership, who may be tempted to imperil their nation's future health in shady transactions that will line their own pockets.

All this militates in favor of banning the shipments outright. And yet, it remains disturbing that many of the nations that appear willing to dispose of the world's wastes are abjectly poor; and it is true, too, and tragic, that uninhabited—even uninhabitable—spaces may be one of their most marketable "resources." Are those who want to ban the shipments being modern Marie Antoinettes, putting off the poor with "Let them make steel"?*

---

* The issues go beyond the ultimate location of an industrial by-product. In an internal World Bank memo leaked to the *Economist*, Lawrence Summers, the Bank's chief economist, invited consideration of the question, "Shouldn't the World Bank be encouraging *more* migration of the dirty industries to the LDCs?" He gave as reasons in support, first, that the costs of pollution are a function of earnings lost through death and wages, and in the LDCs laborer earnings are lowest, therefore "the logic of dumping a load of toxic waste in the lowest wage country is impeccable." Second, in general the damage a polluting agent causes rises with the amount of pollution—low levels can be ignored where concentrated levels can be lethal—"so polluting the cleanest parts of the world may be less harmful than making the dirty parts still filthier." Third, people in general tend to value a clean environment more, that is, "trade off" more basics like housing for environmental amenities, as their incomes rise. Therefore, if there were a migration of dirtier industries from the industrialized North to the LDCs, it would not only add income in the poor countries, it would reduce the global costs of pollution. See "The Freedom to be Dirtier than the Rest," *The Economist*, May 30, 1992, A Survey of the Global Environment, p. 7.

The world is going to continue to produce hazardous wastes. Some of the safest spots to store them may be in the Third World. The quandary is awful, any way one looks at it. The best solution may be not to ban the transactions entirely, but to add another level of protection against the "agency" problem referred to, the sell-out by the self-benefiting dictator. One improvement would be to go beyond the present proviso, which disavows the shipment only if the shipper "has reason to believe that the wastes in question will not be managed in an environmentally sound manner." We should put an affirmative obligation on the shipper to assure itself that the proposed site meets the prevailing standards for containment in the developed world and that the shipper has, after documented inquiry, no reason to suspect that damage to humans or the environment is likely.

A second improvement would be to establish an international authority with power to police waste sites effectively. The organization would have power to preinspect any proposed burial site whose integrity it suspects and to challenge an arrangement if the site is determined to be hazardous. Indeed, any such agency might well be given some power of continual monitoring.[34]

SUMMARY

Transboundary pollution presents a host of enormously complex issues for those who would control it by litigation, or even by diplomacy. Many of the legal devices we array against domestic polluters, such as punitive damages and criminal fines, much less jailing, are practically unavailable in the international arena. Even the injunction, a principal bulwark of domestic law enforcement (since it can be used to restrain damage directly before it occurs) has no counterpart in the scanty arsenal of international practice. Customary principles of international law provide a threatened nation with little legal recourse beyond suing its neighbors for damages after they have already occurred. And even then, as we have seen, the practical influence of such suits is limited by various jurisdictional and doctrinal problems: jurisdiction is noncompulsory, trials are complex and time-consuming, and recovery is, in all events, uncertain. Irreversible damage may proceed far faster than the legal system that is pursuing it.

None of these remarks is intended to dismiss the value of contentious litigation, either in municipal courts, subject to the forum's own law or customary international law, or in international forums, under principles of customary international law. It is not unrealistic to hope that municipal forums will become more receptive to suits by foreign nationals coming to them with transboundary complaints.[35] One may be hopeful, too, that the customary principles can eventually be forged into an instrument to provide compensation for abrupt incidents that can be traced unambiguously to a definite point of origin and associated, perhaps, with an ultrahazardous activity, such as operation of a nuclear plant.

Unfortunately, however, most environmental degradation a nation suffers from "outside" activities occurs too imperceptibly, too "innocently," too ubiquitously and from too many point sources to be stanched by the principles that existing law, domestic or international, makes available to potential transboundary plaintiffs.

For effective measures, we shall have to look elsewhere.

Managing the Global Commons

Much of the most serious damage to the biosphere falls on the global commons—the high seas and atmosphere, in particular.* In some cases, the injuries result from activities that take place within sovereign territory but whose influence extends upward and outward into the atmosphere and high seas; these are the *State-to-Commons* cases. In another group, the *Commons-to-Commons* cases, typified by overfishing and ocean incineration, both the activities and the injuries take place on the commons. In the *Commons-to-State* situations, activities conducted on the commons, such as testing weapons in the atmosphere or dumping ship wastes at sea, have effects that directly and quickly migrate out of the commons to invade areas within national jurisdiction.

If the pronouncements of the most abstractly authoritative legal doctrine are to be believed, there is no legal distinction between these cases and the transboundary cases we have already discussed. The same general principles that admonish nations not to cause unreasonable harm to one another through transboundary pollution typically reprove harming the commons areas in the same voice.[1] But in practical fact and in diplomatic expectation, each of the three com-

---

* Harm to the commons areas may ultimately cause damage within national territories. For example, overloading the atmosphere with greenhouse gases may eventually lead to the submerging of an island nation. But as suggested in chapter 3, from the island nation's point of view the peril is too remote and contingent, proof too complex, and the potential defendants too indefinite a class to expect it to be able to fashion legal and diplomatic relief out of customary principles of international law. Hence, focusing on degradation of commons areas, as such, has the virtue not only of protecting commons areas for their own sake and that of the community of nations, but can be a prudent means of heading off damage that eventually *would* fall within some nation's jurisdiction before the conditions have ripened to the point of perhaps irreparable harm.

71

mons-implicating situations presents distinct challenges to those favoring extended protection. I will deal with each of the three cases in turn.

A Nation's "Internal" Activities
Damaging the Commons

While some degradation of the commons areas can be ascribed to activities that take place directly on the commons, such as oil spills from tankers and incineration of toxic materials on the high seas (see below), the bulk of the invasion consists in waste that originates within national territories, and is carried upward or out across the commons by currents of water and air.

The scale of waste that travels the state-to-commons pathway is gargantuan. Together, the nations of the world pump over 7 billion tons of carbon into the atmosphere annually (the U.S. share is over 1 billion tons), together with 255 million tons of nitrous oxides and sulfur dioxide, and 770,000 tons of CFCs.[2] Similarly, 85 to 90 percent of ocean pollution originates on land.[3] The United States alone disposes in its coastal waters over 60 million tons of dredged materials, 7 million tons of sewage sludge, and 675 thousand tons of industrial wastes, along with trillions of gallons of liquid wastes, and with it, every year, a swelling flotilla of nonbiodegradable trash—plastic bottles, soft-drink containers, and the like.[4] As a result of ocean dynamics and simple gravity, some considerable portion of this waste material is eventually destined to wind up in the deep sea and seabed.

What distinguishes the cases involving harm to the commons (the state-to-commons and commons-to-commons cases) is that the degraded area lies outside any conventionally recognized jurisdiction. The "no man's land" feature of the commons has distinct implications. To begin with, it introduces distinct problems of monitoring. In the transboundary cases discussed earlier, if Canadian fumes are wafting into the United States, there are presumably local U.S. officials on hand to observe the damage over time. The authorities have an opportunity to determine where the fumes are coming from and, if the damage is serious, to institute appropriate diplomatic and legal

relief. But when the commons are imperiled, things are more complicated. Are significant loadings of heavy metals working their way
into the deep seabed, and if so, are the levels dangerous, and who is
responsible? To answer these questions about the seabed, even to
gather the relevant facts, will typically take some special motivation
and effort—perhaps even a multinationally coordinated effort.

Even if the monitors identify substantial and worrisome changes
in the environment and pin down their source, there are complex
obstacles facing anyone seeking to raise a legal challenge. To illustrate, suppose that a nation with an appreciable interest in the ocean
were to underwrite the costs of stationing a ship to monitor deep sea
pollution, and that the ship were to discover that the plumes from a
coastal state's sewage, traced seaward, were imperiling the food
chain of living resources beyond national jurisdiction. What could
the monitoring nations do to stop it?

The first obstacle would be one of finding a legal interest that the
polluting nation was invading. By themselves, *changes* in the quality
of the commons do not necessarily constitute the sort of legal *damage*
that would support a strong diplomatic protest, much less a law
action. Classic principles of international law incline to conceive
"damage" in terms of affronts to sovereignty. When an out-of-control
Soviet satellite fell on Canadian soil, Canada had legal grounds to
complain, demand an explanation, and insist on greater care in the
future, even though the debris fell in an isolated area where it did no
measurable "real" harm. But if Canada discovered that Russian effluents were having adverse effects on Canadian interests in the Arctic beyond Canada's pollution control zone (see below), Canada's
grounds for redress would not be so clear.

A fishing nation might show that it had customarily fished the
areas involved, and that any impairment of the food chain was an
indirect threat to an economic interest: its expectancies of future fish.
There is some chance that such a tack might succeed, if the threat
were clear and high.[5] But the argument is at best uncertain. If someone should come onto your yard, and shoot your pet turtle, you
would have a suit because it would be a trespass and injury to your
property. But if someone shoots a turtle on the high seas, not yet

73

reduced to anyone's ownership, the law is unlikely to grant a remedy. For one thing, the law could not know for certain that the plaintiff would have captured it, rather than someone else or no one at all.

There are in fact scraps of legal doctrine from which some nation might construct an argument that it had a right to sue a country that was degrading the commons areas. The argument would be based upon the view that the commons were not *unowned*, but rather owned *in common* by all nations. The challenger would have to breathe new life into some hoary legal concepts with offputting Latin names: the *actio popularis* and wrongs *erga omnes.** However, no claim arising out of commons despoliation has yet to be pressed in national or international courts, and, aside perhaps from special circumstances where the complaining nation might show that the wrongdoer violated an express agreement (such as a treaty), its prospects would have to be regarded at present as rather doubtful.[6]

In light of the growing concern over the global environment, this gap in the legal system has not passed unnoticed. In its Draft Convention on State Responsibility, the International Law Commission has proposed that "a serious breach of an international obligation of essential importance for the safeguarding and preservation of the human environment, such as those prohibiting massive pollution of the atmosphere or of the seas" be made an international crime.[7] Unfortunately, it is not clear whether any of the existing or presently foreseeable rules would qualify as the essential "international obligation" whose breach would thereby be a crime. Nor is it clear whether any effective sanctions would be levied against the criminal.[8] And in all events, the proposal appears to have made little diplomatic headway since it was floated in 1976.[9]

Hence, if one looks behind the various declarations, principles, and proposals and examines the prevailing practices—the law in action—one sees that, aside from a few areas provided for by special

---

* The *actio popularis* refers to an action that can be brought by any nation, the international counterpart of the domestic "citizen's suit"; wrongs *erga omnes* refer to offenses against the community of nations, a concept historically invoked to legitimate any nation punishing a pirate.

treaty, the commons areas are essentially underprotected. Just as the commons are unowned for purposes of wealth exploitation—anyone can seize a favorable satellite position or scoop up deep seabed minerals without answering to the world community[10]—so, too, no one is really having to answer for their pollution of the commons. The situation is destined to get even worse the more the advanced nations run out of low-cost waste disposal sites on land. What can be done?

The Enclosure Movement

To begin with, one should not disregard the contribution that states can make through unilateral action: each state can rectify trends in the volume of its own transfrontier waste[11] or, at least within the constraints of GATT (chapter 2), apply some of its own domestic environmental laws to activities with transfrontier impact.[12]

But thus far, at least in regard to ocean management, the most dramatic response has been a series of rather bold moves by the coastal states to extend their territorial jurisdiction seaward.

The enclosure movement did not originate as a conservation or environmental measure. It began with an eye on resource exploitation (and subsequently drew additional support from considerations of naval advantage). In 1945 President Harry Truman simply proclaimed that the United States was asserting exclusive rights to exploit the mineral and hydrocarbon resources lying on and under the nation's continental shelf, even where the shelf extended well beyond the traditional 3 miles of territorial sea (whose traditional significance for ship transit Truman, in fact, continued to recognize). Because Truman's proclamation met little resistance, other coastal states began to make further and more extensive claims, typically to the fish and to the mineral wealth of the submerged lands for 200 miles seaward, irrespective of the reach of the continental shelf.

At present, 144 coastal states have asserted seaward territorial extensions,* with the result that the most resource-rich third of what

---

* Out of 150 coastal states globally; the position of the other six (which includes new states such as Bosnia-Herzegovina) is not definitive.

was formerly open ocean has been enclosed. Put another way, the ocean portion of the global commons has shrunk 30 to 40 percent assuming, as now appears to be the case, that the jurisdictional extensions, the so-called Exclusive Economic Zones, will be recognized.[13]

From the perspective of the global environment, the enclosure movement has two important implications. First, before enclosure, any state's authority to control damages in the "open" area beyond the traditional 3 or 12 miles was, as we have seen, totally unclear. With enclosure, a broader domain at least receives a governor: the enclosing state assumes the competence to manage the area much as though it were part of its land mass. Second, with this "privatization" come incentives to manage *well*. The coastal state will be circumspect about how the region is governed, because the costs of abuse no longer fall on the community of nations as an undifferentiated whole, but are concentrated on the coastal state itself, as "owner."

This division of the oceans has defenders, most persuasively in the case of living resources such as commercially valuable fish. Because the living resources regenerate, their production over time can be maximized as long as the rate at which they are taken does not exceed a biologically determined optimum for the species and habitat. As we saw, too, so long as the ocean areas are commons and no one "owns" the fish until they are netted, the economic and biological ideals are virtually unachievable: each fishing nation fears that if it does not net the fish, someone else will—leading to the so-called Tragedy of the Commons.

On the theory that the fish and other ocean resources are the common heritage of humankind, the offshore areas might have been consigned to the world community to manage, as many commentators advocated. But getting the splintered "world community" to agree on an international management regime for the oceans was part of the agenda of the Law of the Sea (LOS) Conferences, and the ambitions have proved elusive after decades of negotiations. The easiest alternative—the path of least resistance—was to accede to the bold claims of the coastal states.[14] The coastal states could point to their special long-term interest in maintaining the stock of fish off their respective shorelines, and to their strategic proximity to act as "policemen."

There is a similar argument for enclosure from the perspective of pollution control. As long as the waters and seabeds beyond the coastal area are someone else's, each coastal state can regard its pollution as (to some extent) "someone else's problem." In other words, the coastal state, in weighing the costs and benefits of pollution abatement (sewage treatment before discharge, etc.), notes that it bears all the costs of the cleanup efforts, but realizes only a fractional share of the benefits. Some of the benefits of treating discharges accrues to those with an interest in the commons, who will be "free-riding" on the coastal state's efforts. Hence, from the coastal state's point of view, as the ocean space over which it has jurisdiction pushes outward, cost/benefits calculations shift in favor of further treatment of its discharges.

Those, at least, have been among the theoretical justifications for the enclosure movement. Several things need be said about this reasoning, however.

First, the argument for enclosure rests largely upon claims of management *efficiency*; it does not respond to the claims of the land-locked and less developed countries that enclosure is unfair on *distributional* grounds: that it is simply unjust for states that happen to have long coastlines (including some of the already richest nations) to bolster their economies by snatching up what many consider to be part of the common heritage of humankind.*

Second, even regarded from an efficiency point of view, there is no convincing evidence that the enclosures have been successful even at their strongest promise: at stabilizing fish stocks. The reporting of landings has almost certainly become more accurate. But enclosure, of itself, does not eliminate overfishing. While it removes the pressure attributable to international competition in the enclosed area, the enclosing state has to follow up the enclosure by restricting exploitation, including activities of its own domestic fleet. Under coastal-state management, the managing states can, and in practice

---

* I will return to this question in chapter 9.

generally do, license and restrain the harvest by outsiders as well as by their own fleets. There is an intriguing, but still incipient prospect of benefits to be realized by one nation selling its quotas to others, including conservationists (chapter 6). But thus far the coastal states, including the United States, have lacked the political will to stabilize stocks. To do so would require controlling their domestic fishing industries. Typically, the exclusion of "foreign" ships has been attended by an offsetting increase in the domestic fleet, which in fact often expands as some of the ousted foreign shipowners have to unload their technologically superior vessels at depressed prices. Whether overfishing will really be remedied by national fisheries management is, at best, still an open question.*

Finally, the advantages of the enclosure movement in regard to pollution have been even less clear, even in theory. It is true that the extension increases the ability of the coastal state to protect its beaches from pollution by foreign vessels.[15] But so far as our concern is with ocean pollution, the primary offender is ordinarily the coastal state itself. In other words, it did not take enclosure to give the coastal states considerable incentive to treat their discharges, since most of the damage was always occurring in the relatively fragile coastal waters that were theirs under traditional 3- and 12-mile limits. Hence, while enclosure should encourage the coastal state to improve pollution control in theory, in practice the actual reduction of waste loads are probably minimal.

The optimist can, however, cite one indirect benefit of the EEZs in the control of ocean pollution. By clarifying jurisdictional issues and reducing the number of nations with competence to govern any area, the movement had the unintended effect of facilitating the efforts of the United Nations Environmental Programme (UNEP) to promote its Regional Seas Program (RSP). The RSP involves fostering a series of

---

* Note, too, that the virtues of coastal state management, so little borne out thus far in the case of fisheries, are even less clear where nonregenerating resources are concerned. Unlike fish and other living resources, offshore oil, gas, and hard minerals do not reproduce, and thus there is less net production advantage in metering the pace of exploitation.

agreements among nations bordering each of several seas to work out mutually acceptable pollution control programs. Although ten agreements have been concluded or are under negotiation (including, at present, the Mediterranean, the Caribbean, and the Persian Gulf), it remains too early to judge how successful they are going to be. At this point, most of the agreements have proceeded to the point of agreeing to study the situations further—unsurprisingly, since the interests that favor pollution are generally powerful, clean-up costs are large, and the RSP managers have few resources at their disposal. Nonetheless, the program is clearly at the groundwork stage, and it at least provides a framework for further action as the climate for meaningful action, hopefully, improves. Indeed, in regard to the Mediterranean, which has the most long-standing of the RSPs (signed in 1976), there is some sentiment that while pollution has not abated, the situation is considerably better than it would have been in absence of the so-called Med Plan.[16]

In summary, enclosure—a sort of privatization on a large, public scale—may have some benefit, yet unproven, in conserving the oceans. But as a general strategy for protection of the other reaches of the commons, it hardly presents a strong reed to lean on.[17]

Special Treaties

Another approach—in the opposite spirit of enclosure—is to negotiate treaties providing for cooperative management of particular subject areas or regions.* Interestingly, while there are several conventions that restrict ocean dumping and pollution from ships (situations that fall under the *Commons-to-Commons* and *Commons-to-State* cases and are thus treated below), aside from some of the RSPs there are few conventions that aim to restrict land-to-sea pollution. There is a promisingly titled Paris Convention for the Prevention of Marine Pollution from Land-based Sources, which was signed by fourteen nations in 1974 and amended in 1986 to rectify the original failure

---

* The UNEP Regional Seas Programs are a hybrid, combining enclosure with co-operation.

to deal with the vast quantities of pollution that reach the sea via the atmosphere.[18] While a commission was once formed under it to create environmental quality standards which would take effect two hundred days after their unanimous adoption, nothing has ever come of it.[19] UNEP has developed guidelines for land-based discharges (known as LBDs), but they are nonbinding.[20]

It is instructive to consider why comparatively little headway has been made in regard to controlling the masses of pollution that travel a land-to-sea pathway. One explanation is the resistance of countries to relinquish autonomy over what they do within their own borders, as compared with what they do on the commons areas (see below). But if internal autonomy were crucial, it would have undermined any number of international treaties, going back to the migratory-animal conventions of the nineteenth century, right through the sulfur dioxide and ozone accords (see below).

It could be thought that because land-based discharges pass through the polluting nation's own territorial waters and air basins in the first instance, we can rely on an appreciable independent motivation for the polluting state to exercise restraint. But it is fairly clear by now that any such hope is misplaced. Agreement may be hindered by lingering scientific doubts as to how harmful marine pollution really is, or perhaps the anticipated difficulties of detecting violations if restrictions were established.

My guess is that the real stumbling block is simply practical economics. The fact that so much waste is involved—and growing—and so many hard-pressed industries and nations are implicated, means that until public protest reaches a shriller pitch, little will change.

THE OZONE CONVENTION

The situation with ozone was and is distinct. The Vienna Convention for the Protection of the Ozone Layer (1985) established a general framework contemplating a series of refinements, which have since been provided by successive ozone protocols (additions) arrived at in Montreal (1987), London (1990), and Copenhagen (1992). At this point, more than eighty countries are coordinating efforts to phase out ozone-depleting substances.

The ozone agreements are not without their critics. Some maintain that, considering the existing inventory and life expectancy of chlorine already injected into the atmosphere and the recent indications of more serious impairment than was first supposed, the phase-out schedule is too slow. Some dangerous agents are presently uncovered. And some of the permitted substitutes have themselves fallen under suspicion.[21] In fact, enthusiasm for the convention now has to face the fact that while the estimated $4 to $6 billion costs required to substitute CFCs (principally with hydrochlorofluorocarbons, or HCFCs) was justified in London on the basis that industry would be allowed to use HCFCs until 2040, already by 1992 Mostafa Tolba, executive director of UNEP, was calling for phasing out HCFCs by 2005. If that is accepted, the HCFCs would have to be replaced largely by other chemicals (principally hydrofluorocarbons, or HFCs) whose production and maintenance costs are higher. Moreover, while HFCs and other agents are chlorine-free and therefore ozone-safe, they present a range of other risks, including flammability, corrosiveness, toxicity, and dangerously high operating pressures.

Most commentators probably feel that, notwithstanding these ordinarily undiscussed blemishes, in the context of international environmental regulation the ozone agreements, in garnering so quickly such firm commitments from so many nations, are an outstanding success. And, inspired by the cooperation the ozone negotiators achieved, it has become common to ask why we cannot achieve the same success with land-to-sea pollutants, or, for that matter, with greenhouse gases.

The answer is that, without detracting from the honors the ozone negotiators deserve, reaching agreement on the ozone shield was in several respects easier than the task that faces those seeking a diplomatic breakthrough in the other areas.[22] It is helpful to keep some of those unique features in mind.

Scientific consensus. By 1987 there was widespread consensus both that the ozone shield was measurably thinning, and that—whatever the true risks might prove to be—the depletion was occurring at a rapid clip. Moreover, there was a subtle corollary to the speed: the public had not had time to grow accustomed to the idea that a thinning ozone shield was something that we all just "had to live with." There is, to be sure, the same sort of general agreement that the ocean is getting dirtier and the blanket of greenhouse gases is thickening. And public awareness of both issues is, fortunately, increasing. But as we saw in chapter 1, there is much less consensus among scientists about whether the earth is warming appreciably or even about what foreseeable warming would bring. There are analysts who say of GHGs what was rarely said about CFCs: that we do not lose, and may even gain, if we defer costly investments in prevention and mitigation until we have a clearer picture of the real risks.[23]

Winners and losers. At the Montreal and London conferences, there were differences of opinion about how bad the perils of ozone thinning were, and where they were likely to fall hardest. But at least no nation could foresee any *benefit* from ozone thinning.[24] By contrast, to protect from greenhouse warming, negotiators will have to work out an accord among nations that may realistically face (or can credibly claim to face) appreciable gains in agricultural productivity and even heating costs that might offset some of the generalized losses that would come of living on a more tempestuous planet.[25]

Number of indispensable parties. In general, the greater the number of parties required to make an accord effective, the more difficult the agreement is to forge. Going into Montreal, only a handful of countries and, for that matter, manufacturing firms—all partners in the industrial world—accounted for virtually all the world's CFCs.[26] By contrast, hammering out an agreement to restrict GHGs involves bringing together virtually every major nation in the world.[27]

Costs. While reliance on CFCs and other ozone-depleting agents had become extensive by the late 1980s, the industry, at its widest expanse, accounted for only $750 million in sales worldwide.[28] And not all their profits were hanging in the balance: many of the same firms that saw their CFC sales jeopardized were preparing to go with substitutes for many uses that were either already at hand or on the horizon. (Some computer chip manufacturers found that CFC rinses could successfully be replaced with soap and water!) And from the industry's vantage point, some shift away from CFCs may have appeared legally prudent; could there have been no discussion in the boardrooms of possible liability to skin cancer victims along the lines of cigarette company liability to cigarette smokers who contract lung cancer? No one can say for certain how the law will change, but, as we saw, the prospect of a GHG emitter being held legally liable for climate change damage seems quite remote. Moreover, adequate quantities of affordable, atmospherically benign substitutes for fossil fuels and methane sources, in the amounts needed, are not in the offing (particularly if nuclear energy remains in disfavor), even crediting the most optimistic estimates of demand reductions that may be achieved through conservation.[29] For the United States alone, the cost estimates for freezing carbon output at 1990 levels through the next century have been estimated to run between $800 billion and $3.6 trillion.[30*] While there are, on this issue, widely varying ranges of optimism and pessimism, it is hard to argue that a virtual *elimination* of all GHGs, or even of CO_2 only, over a period of ten or fifteen years is realistically attainable.[31]

This is not to say that measures to reduce ocean pollution and greenhouse gases ought not to be sought; only, that the apparent success of the ozone negotiations ought not to be generalized into high hopes for comparable accords in other areas.

Guardianships

To be counted a success, an arrangement does not have to achieve a full and immediate phaseout of some risk-creating activity. Subtler and less sweeping measures ordinarily must be allied, as called for by the circumstances.

---

* Even if enough other nations were to go along so that our billions or trillions of dollars succeeded in avoiding a doubling of atmospheric CO_2, it is not clear whether the gains would be worth it. William Nordhaus points out that the great bulk of the U.S. economy is unaffected by the weather outside; he estimates that climate changes associated with a doubling of CO_2 would reduce U.S. national income by only 0.25 percent. William D. Nordhaus, "To Slow or Not to Slow: The Economics of the Greenhouse Effect," *Economic Journal* 101 (1991): 920 and table 6.

As we saw, one of the reasons for overexploitation of the commons is the lack of a plaintiff clearly qualified to demonstrate both standing and injury. The antidote I have been proposing is a system of guardians who would be legal representatives for the natural environment. The idea is similar to the concept of guardians (sometimes "conservators") in familiar legal systems. Presented with possible invasions of the interests of certain persons who are unable to speak for themselves, such as otherwise unrepresented infants, the insane, and the senile, courts are empowered to appoint a legal a guardian to speak for them. So, too, guardians can be designated to be the legal voice for the otherwise voiceless environment: the whales, the dolphins, important habitats, and so on.[32] The guardians could either be drawn from existing international agencies that have the appropriate focus, such as the World Meteorological Organization for the atmosphere, or from the many nongovernmental organizations (NGOs) that might be willing to serve, such as the World Conservation Union and the World Wide Fund for Nature (WWF).

A guardian would not have the exclusive power to call a halt to a threatening activity. Rather, the guardian would be inlaid into the lawmaking and adjudicative process. For example, to assure that oceanic ecosystems were being adequately accounted for, an ocean guardian might be designated, perhaps GESAMP, the Joint Group of Experts on the Scientific Aspects of Marine Pollution, with supplementary legal staffing. The guardian would be authorized (1) to monitor ocean conditions, (2) to appear before the legislatures and administrative agencies of states considering ocean-impacting actions to counsel moderation on behalf of its "client"; (3) to appear as a special intervenor-counsel for the unrepresented "victim" in a variety of bilateral and multilateral disputes. Perhaps most important (4), international treaties might endow the guardian with standing to initiate legal and diplomatic action on the ocean ecosystem's behalf in appropriate situations—to sue at least in those cases where, if the ocean were a sovereign state, the law would afford the state some prospect of relief.

The notion is hardly far-fetched. Indeed, many guardianship functions are currently recognized in U.S. environmental laws on a more modest scale. For example, under the Superfund Legislation, the National Oceanic and Atmospheric Administration (NOAA) is the designated trustee for fish, marine mammals, and their supporting ecosystems within the U.S. fisheries zone. NOAA has authority (through application to the U.S. attorney-general) to institute suits to recover restoration costs against any party that injures its "ward."[33] A major lawsuit is now proceeding in federal court in southern California, in which NOAA attorneys are suing local chemical companies allegedly responsible for seepage of PCBs and other chemicals into the coastal water ecosystem.[34]

There has been a recent case in Germany that actually invoked the guardianship concept in the global commons context. In 1988 approximately 15,000 dead seals mysteriously washed up on the beaches of the North and Baltic seas. Widespread alarms were sounded amid considerable concern that the massive deaths (which the German press dubbed *Robbensterben*—"seal death"—to echo the long-publicized *Waldsterben*) were a portent of an impending ecological disaster. Many agents fell under suspicion, but at the time, the most flagrant insult to the North Sea's chemistry was widely considered to be the titanium and other heavy metals, by-products of a West German lead and paint complex, that were being incinerated and dumped on the high seas by permit of the West German government.[35]

Conceivably, any of the states bordering the sea might have tried to challenge Germany's actions. But recall that, so long as the harm was being done on, or affecting life only in, the high seas, the authority of any nation to sue was (and is) doubtful. For Poland, say, to trace through a compensable injury would have been nearly hopeless. From the point of view of national fishing interests, the reduction—even elimination—of the seals (themselves commercially valueless but voracious) might even have been regarded as an economic *benefit*. Moreover, *all* the littoral nations were contributing to the pollution, and thus, had any of them brought suit, their case might have been met by Germany with what lawyers call an "unclean hands" defense: "You can't complain, because you're as guilty as we are."

85

Who was there, then, to speak for the seals, and, in so doing, represent all the elements of the ecological web whose hazarded fortunes were intertwined? In comparable situations in the United States, courts have shown willingness to interpret the Administrative Procedure Act and other laws as giving a public interest group standing to challenge the government's actions.[36] German law, however, is not comparably primed to allow "citizens' suits" to take government agencies to court.[37]

The solution was for a group of German environmental lawyers (with my encouragement and advice) to institute an action in which the North Sea seals were named the lawsuit's principal plaintiffs, with the lawyers appearing essentially as guardians, speaking for them. Hence the case name: *Seehunde v. Bundesrepublik Deutschland*.[38] And what better plaintiffs? No one could accuse the seals, surely, of unclean hands (or fins). And the injury to them was immediate, rather than remote and conjectural like the harm that the other littoral nations, such as Denmark, would have had to rely on had they chosen to sue.

The administrative law court rejected the seals' standing (because seals are not "persons" and no specific legislation authorized standing on their behalf) and dismissed the action.[39] But the outcome was doubly instructive. First, it showed you could win even while losing. The very filing of the case and attendant publicity was considerable and favorable.[40] When the time came for the government to renew the ocean-dumping permit, the authorities who initially gave their permission were forced by a kindled public opinion to revoke it. Germany has committed to phase out the practice of disposing heavy metals into the North Sea. The seals lost the battle in court, but even so gained an advantage in ultimately winning the war in the public and political forums.

Second, the rejection of standing in the name of the *Seehunde* themselves underscored the fact that for an environmental group to seek guardianship representation on an ad hoc basis is an insecure strategy. Without special advance authorization, the chances of a formal body recognizing the authority of someone who steps forward to speak on behalf of a natural object (the species or whatever) are not large, not even in U.S. courts, which have been the most receptive, at least until recently, to widen the courtroom doors.*

The preferable route—special advance authorization institutionalizing a guardianship system for the global commons—would almost certainly require international action, including formal recognition by appropriate international bodies of the guardianship connection. But there is no reason why the desirable approval could not be arranged. While the guardians' standing before the World Court would not be indispensable (because there are many forums in which the guardian might appear), such standing could be secured by amending the charters of the United Nations and of the World Court.[41]

The guardianship approach has one distinct advantage over the nearest alternatives, revival of the *actio popularis* and *erga omnes* doctrines referred to above. Under those approaches, any state that wanted to complain would have the power to initiate an action. There could thus be many conflicting suits and judgments in courts here and there, involving the same sea or species. Some judgments might be inconsistent and even if not, the prospect of litigation might never be set to rest. If Poland sued Germany over pollution of the Baltic and Germany were vindicated, Germany might still have to defend itself a second and third time in subsequent Baltic-centered

---

* I summarize the fate of a number of such cases in my *Earth and Other Ethics*, pp. 6–9. For example, in 1976 several animal welfare groups sued the secretary of commerce to halt the issuance of permits to import skins of South African fur seals taken in an inhumane manner. To meet the standing requirement, the plaintiffs claimed that they had an interest in the maintenance of a safe, healthful environment for marine mammals. The district court rejected their claim on the grounds not only that South Africa was remote from the plaintiffs' homes, but that the area of the Cape where the seals were being slaughtered was accessible only by special permit of the South African government—permits not likely to be granted to seal watchers! On review of the district court's judgment, despite the thinness of the mammalian groups' supposed "own interests"—they were transparently suing on behalf of the seals in all but name—the U.S. Court of Appeals reversed and allowed the suit to go forward. Animal Welfare Institute v. Kreps, *Environmental Law Reporter* 7: 20,617 (D.C. Cir. 1977). It is the highly liberalized standing permitted in *Kreps* that Justice Anthony Scalia intended to cast doubts upon in the *Lujan* case discussed in chapter 3.

actions brought by England and Denmark, etc., each of which thought that it could put on a case with a slightly better wrinkle. There being no designated plaintiff uniquely authorized to maintain the action, it would be more difficult to obtain a final judgment.

The guardianship system, by designating one responsible voice for each part of the environment, thus offers a distinct advantage. There is, however, at least one drawback that grows out of that virtue. The more power a guardian were to have, and the more exclusively its voice were the voice that counted, the greater would be the political pressures to compromise its scientific and legal integrity.

Furthermore, while a system of commons guardians would be a step forward, it would be no panacea for biosphere degradation. Those commons areas that were placed under guardianships, such as living ocean resources, would be elevated to a legal and diplomatic standing on a par with a sovereign state. But we have already seen that under present law the position even of sovereigns has only limited advantage in protecting commons areas. Hence, the success of a guardianship regime would depend not only upon legitimation and institutionalization of the guardians, but upon significant changes in the substantive law that the guardian would be empowered to invoke. The oceans not only need their own independent voice; they need some more sympathetic standards.

A Nation's Activities on the Commons Damaging the Commons

The fourth situation involves restrictions on activities deleterious to the commons that are conducted not on a nation's own territory, but in the commons areas themselves. Examples include weapons testing and other scientific experiments in space, dumping and incineration of wastes on the high seas, exploration and mining on the seabeds and Antarctic regions, and overexploitation of the ocean's living resources through modern, potentially devastating fishing technology. The National Academy of Sciences estimates that ocean sources dispose 6.4 million tons of trash into the marine environment annually, most of which is probably innocuous, but some of which, including 45,000 tons of plastic debris, is not. In fact, as far as litter is concerned, land-originating plastic pollution, as bad as it is, is less onerous than the debris that is dumped by vessels and offshore facilities, which combines in the sea with tons of lost and abandoned nets and other fishing gear. Today these things are made predominantly out of plastic, and thus, unlike hemp, are fated to resist decay for 250 to 400 years.[42] Together, it is responsible for a holocaust of birds and sea creatures that die from entangling in or ingesting plastic debris.[43]

Clearly, there is no different range of regulatory *strategies* applicable in this class of cases. Environmental goals could be advanced by a system of guardians; pollution could be abated (and funds for global protection obtained) by charging fees or auctioning permits for deep-sea dumping;[44] there are numerous opportunities for technological mandates, such as seine size requirements.

But there are two differences when we move into the commons-to-commons area. One distinction suggests why regulation of these cases may prove to be harder than in the nations-to-commons cases; the other suggests at least one reason why regulation may meet somewhat less resistance.

The special challenge owes to the sheer vastness of the commons areas. Historically, it has been hard enough policing ships for marine refuse when the ships are in harbor and coastal waters. On the high seas, any policing other than self-policing is virtually futile.

The prospects of monitoring compliance with fishing regulations are not much better. The United Nations has passed a nonbinding resolution requesting all the nations of the earth to cease drift-netting. Japan (as part of a welcome program to lift its status in the world community) as well as Taiwan and South Korea have indicated that they will go along with it by the end of 1992.[45] But a reduction in drift-netting operations may simply lead to a higher intensity of non-drift-net fishing—with the same destruction. Indeed, many commentators contend that while drift-netting has provided a dramatic focus on the issue of fishing technology, it actually yields a lower rate of bycatch than other techniques over which the countries are negotiating, including midwater trawling, shrimp trawling, and longline fishing. Environmentalists have long complained that shrimp fishing in the Gulf of Mexico, not reliant on drift-nets, is the most atrocious of all.

In all events, the success of any fishing "ban" depends ultimately on monitoring compliance. As in all maritime regulation, there is the risk that some nominally "cooperating" nations will allow their ships to shift registry ("reflag") to a third nation, or even look aside as their ships operate out of home ports in violation of announced commitments. The fact is, the world cannot put an inspector on every deep-sea fishing vessel; inspectors bold enough to go on board have all the influence of pariahs. And considering that two thirds of the planet's surface is ocean, just keeping tabs on the drift-netters' whereabouts is bound to prove elusive.*

The same vastness potentially hampers any plan to patrol for ocean disposal of toxic wastes. It is one thing to monitor stationary coastline point sources, such as rivers and sewer outlets, and quite another to monitor ocean dischargers, who are always on the move. It is true that the volume of wastes that flows from fixed-point coastal sources is considerably larger. But the material that is dumped on the ocean, albeit in far more modest quantities, tends by the very nature of the practice (the added costs the dumpers are accepting) to include a disproportionate share of highly lethal refuse, for example, canisters of radioactive waste. Indeed, should the international community require fees or tradable permits for ocean disposal, the higher the price it charges, the greater would be the temptation to dump illicitly.

But the prospects for commons regulation is not all bleak. Technology may be able to provide some assistance in overcoming the

---

* The difficulty of keeping tabs on international ship movements is underscored by the U.S. Navy's "loss" of a North Korean ship the Navy was trying to track in March 1992, because of fear that the ship was carrying contraband Scud missiles to the Middle East. The U.S. assigned a destroyer, a five-ship carrier battle group, and a U-2 surveillance airplane near the Gulf of Oman, just outside the Persian Gulf—to no avail. On the other hand, tracking vessels under a cooperative fishing agreement, would be easier than tracking military vessels, if the nations could agree to having ships fitted with tamper-proof radio equipment that transmitted location.

monitoring problems. For example, to help keep track of ships' whereabouts, we can require vessels to be outfitted with radio-transmitting devices. If manufacturers were required by law to manufacture plastics with biodegradable material (or even if ships were required to use only biodegradable plastics on board), inspection for compliance would thereby be shifted away from the open sea to factories or ports.

Moreover, the promising feature of the situation is that nations are readier to recognize the legitimacy of international control over their activities in space or on the high seas than in their own territories. This is true both because such arrangements make no concessions regarding traditional autonomy over internal affairs, and because all parties know that without such agreements, the unmanaged commons are especially susceptible to abuse.

As a result, as far back as 1911 the United States, Great Britain, Russia, and Japan had reached agreement on the Preservation and Protection of Fur Seals in the North Pacific and Arctic region.[46] While, as we saw, control over pollution of the sea from land-based sources remains difficult to negotiate, there are several accords on marine pollution going back to the 1954 Convention for the Prevention of Pollution of the Seas by Oil. Marpol, the Marine Pollution Convention, has recently been amended to outlaw the dumping of plastics at sea.[47] The International Whaling Convention (IWC, 1935), originally promoted by the whaling nations to maximize whale catches over time, has become increasingly responsive to the influence of "conservationists" in the strong sense of being more sympathetic to conserve whales than whalers. Consequently a freeze on whaling has been in effect since 1986.* The point to underscore is

---

* The moratorium was originally announced as temporary until three tasks could be accomplished: (1) depleted stocks would be provided a "breather" to recover, (2) scientific research to produce a better picture of the whales' numbers and vitality could proceed, and (3) a new "management" procedure could be submitted that would avoid overexploitation. By the time of the 44th Annual Meeting, however, in June 1992, a number of nations had come to treat the moratorium as a permanent ban, shifting focus from "sustainability" to protecting whales for noncommercial, essentially moral reasons. The whaling states feel they have been deceived and aban-

that the whales—a common "resource"—have come to enjoy a level of protection that the black rhino and similar endangered creatures, the "resources" of various individual nations, are denied.

There are other reasons why regulation of other of the commons-to-commons problems may be relatively easy to secure, depending on some particulars. Consider the "successful" conclusion of several international treaties regarding space, for example, the 1967 Treaty on Principles Governing the Activities of States in the Exploration and Use of Outer Space, including the Moon and other Celestial Bodies. The ease of agreement undoubtedly stems from the fact that setting the ground rules for space exploitation does not threaten any established economies, even producers, as might requirements for retrofitting chemical plants and utilities to bring them up to mandatory levels of effluent emission. Moreover, so few nations have the technical competence to affect outer space that the number of signatories required to make the agreements effective was, practically speaking, small.

New, unanticipated problems in the commons-to-commons category arise all the time. In 1991 a controversy erupted between two environmental camps. A group of physical oceanographers interested in measuring changes in ocean temperature, a vital parameter in global-warming studies, proposed to fire pulses of noise through loudspeakers in the Indian Ocean and analyze the sonic blasts as they traveled around the world's oceans. When marine mammologists learned about the pending tests, they were aghast,

---

doned. Iceland and Norway (even under the administration of Gro Harlem Brundtland, a sometimes heroine of the environmental movement) have dropped out of the IWC and will resume hunting. France is floating the idea of a whale sanctuary in the southern Atlantic Ocean up to 40 degrees latitude—a zone that would infringe several Southern Hemisphere EEZs. A three-fourths majority is required to overturn the ban. The U.S. is prepared to continue the moratorium; at least the Bush administration will not initiate action to overturn it. Japan is clearly uncomfortable with the situation but is stopping short of withdrawal, presumably to avoid added environmentalist heat. The next move will be at the 1993 meeting to be held in Tokyo.

fearing that such noises might disorient, if not deafen, whales and other cetaceans. The mammologists were particularly aggrieved because they had never been consulted and thought that the approval had been shuffled all too quickly before U.S. review agencies.[48]

For me, the interesting question is the implicit one: Why should such an experiment affecting the world's oceans be considered a matter for *American* scientists and agencies to resolve? A suggestive illustration of how the community of nations might respond is afforded in the area of scientific tests in space. In the 1960s the United States engaged in an experiment called Project West Ford designed to scatter millions of copper needles 2,000 miles up in space in order to determine their potential as a basis for a communications network, as an alternative to launching satellites. There was a considerable outcry from astronomers around the world who feared deleterious effects on radio astronomy. While the fears have since largely subsided,[49] the protests were influential in getting the International Council of Scientific Unions to establish a Consultative Group on Potentially Harmful Effects of Space Experiments (COSPAR-CG). COSPAR's report, ultimately adopted by the United Nations Committee on the Peaceful Uses of Outer Space, recommended that future comparable experiments should be evaluated by the scientific community before implementation.[50] Moreover, in a movement boosted by the same West Ford episode, the U.N. General Assembly passed a declaration calling for states to "undertake appropriate international consultation" before beginning an experiment that may interfere with the peaceful use or exploration of space.[51] Similarly, the 1967 Outer Space Treaty invites any signatory who is concerned about another signatory's behavior to request information and consultation.[52]

Thus, there is presently a norm—some basis for an expectation in the world community—that nations planning to undertake activities in space will consult an international committee of respected experts in advance. It is hard to imagine why the same principle ought not to apply to potentially harmful activities in ocean space.

A NATION'S ACTIVITIES ON THE COMMONS
DAMAGING ANOTHER NATION

Finally, many activities that take place on the commons pose a threat to national territories: beach-invading pollution from discharges on the high seas, or debris that tumbles from an earth-orbiting satellite. As a matter of international law, the fact that the harm-causing party acted out on the global commons is no reason to relieve it of accountability. In fact, international sentiment tacitly regards these cases as a more legitimate subject of regulation than their mirror image, the state-to-commons situations, in which the effect of activities within national territories are felt on the commons.

For example, the 1971 International Convention for the Establishment of an International Fund for Compensation for Oil Pollution Damage provides compensation for damage suffered on the territory, or within the territorial sea, of a nation. But damage to the seas or sea life from the same spill is not covered. Incursions from outer space into national territory are likewise treated severely. The 1963 Treaty Banning Testing of Nuclear Weapons in the Atmosphere, in Outer Space, and Under Water has to be regarded, of course, as part of the arms negotiation package, but the protection of populations from fallout is also a principal motivation.[53] There is a 1972 Convention on International Liability for Damage Caused by Space Objects, which subjects the nation whose vehicle is responsible to strict liability, that is, liability without requirement that the injured state demonstrate negligence.[54] The Convention actually came into play in 1978 when the Soviet *Cosmos 954* military satellite crashed in northern Canada.[55]

Of course, compensating damage that has already occurred is not as satisfactory as averting the harm in the first place. One preventive strategy in the commons-to-state area is represented by Canada's response to threats of pollution to its fragile Arctic frontier. Canada in 1970 declared a unique antipollution zone stretching 100 miles north of its coast (a brazen move, considering that, at the time, the 12-mile limit of the 1958 Geneva Convention was the general rule) in which it claims special limited jurisdiction to regulate all potentially polluting activities.[56] The law (the Arctic Waters Pollution Prevention Act) goes so far as to claim power to seize ships that operate in the antipollution zone in violation of Canada's regulations for the area.

Canada's legislation remains controversial, largely because the purity of its motives and evenhandedness of application are in question. Its antipollution zone would seem to deserve support among bona fide environmentalists. But there is an uneasy suspicion that the Canadians intended to secure their access to whatever petroleum and mineral wealth may lie in the far north by approving only their own exploration ships and facilities as environmentally sound, and fending off other nations with the threat of being tied up in protracted legal proceedings. The prevalence of these misgivings, however justified, may provide further reason to press for stronger international regulation over commons area pollution. But in the interim, without effective international control of activities on the commons that endanger states, some self-help by the coastal states, particularly if subject to international guidelines, cannot be regarded as unreasonable.

Indeed, many nations, including the United States, claim the right to exercise extended seaward claims as far as is required to keep "foreign" fishing vessels away from "our" anadromous fish on the high seas.* Anadromous fish, such as salmon, spawn and are traditionally captured in inland rivers but live most of their lives in the high seas, where they are vulnerable to international fishing fleets. Certainly it seems more defensible for a coastal state to protect itself by an extended pollution or anadromous fish zone than to arrogate to itself the wealth of the EEZs without providing any compensation to the landlocked and less developed countries.

* In 1992 the potential for conflict was narrowed when Canada, Japan, the Russian Federation, and the U.S. signed a convention agreeing to halt all salmon fishing in the high seas of the North Pacific Ocean.

CHAPTER V

Treaties as Antidotes

THE VIRTUES OF A TREATY-BASED APPROACH

I HAVE expressed doubts that the slow-growing body of general principles of international law (those rules that prevail in the absence of any special treaty) should be relied upon to provide significant protection for the biosphere. Part of the explanation lies in the list of hurdles a litigating nation faces: getting the defendant to submit to jurisdiction, resolving complicated questions about causality, meeting standards of state responsibility, obtaining satisfaction on a judgment, should one eventually be forthcoming, and so on.

All of these defects in the general rules point to the advantage of bilateral and multilateral treaties that are hammered out and specially tailored to anticipate specific disputes in advance.[1] In conference, the negotiators can translate the nebulous contours of the customary law (which often amount to little more than "don't use your territory in such a way as to injure others") into a more specific, reckonable and realistic set of obligations. Although treaties bind only the signatories, which is a distinct limitation, the other side of that coin is that the signatories have control over whom they are going to deal with, and who will be empowered to call them to account.*

There are several virtues to such agreements. First, while customary international law moves slowly, treaty negotiations provide focal opportunities both to clarify the controlling law and to establish the

---

* In addition, the more signatories a treaty can marshall, the stronger the case for applying the principles it embodies to nonsignatories indirectly on the basis that the widespread acceptance of the treaty is evidence that its principles have become part of the body of generally accepted principles and "customary law" that binds all members of the world community.

governing procedures. For example, treaties can flexibly specify which sorts of disputes will lead to diplomatic response, and which will be turned over to arbitrators or to courts for binding third-party disposition. The International Court of Justice (ICJ) can be given compulsory jurisdiction if the signatories so desire.[2]

There is a medley of other alternatives. Under a 1974 environmental accord, the Nordic states (Denmark, Finland, Norway, and Sweden) agree that each of them, in permitting a potentially harmful activity, has to account for the interests of the other states that may be affected. And then, to solidify the obligation, the convention empowers any citizen of a contracting state (and not merely an official) who is or may be affected by an emanating nuisance to institute proceedings before the appropriate forum of the offending state. In other words, if a Swede is aggrieved by actions taking place in Denmark, and for some reason of diplomacy the Swedish public representatives will not press a claim on the citizen's behalf, the individual Swede can take her cause into Denmark herself—and the Danish courts and agencies have to give her standing.

Under treaties, the distinctive problems of assuring protection for the commons areas (see chapter 4) can be provided for and, to some extent, overcome. Moreover, treaties characteristically adopt a preventive approach, reducing or eliminating risks before they have ripened into legally provable injuries. The ozone agreements display this preventive emphasis. Whatever compliance measures are finally agreed to, if any (the issue of sanctions has been deferred to subsequent negotiations), the duties are spelled out in terms of *usage*, not *harm*. This means that a complaining nation need not demonstrate actual injury, much less prove a causal link between that injury and the violator's CFC emissions. A nation is in violation of the agreement once it has exceeded its allowable level of use, period.

In addition, if a nation has reservations about putting itself under externally imposed hard and fast "laws," treaties can provide for international agencies that can promulgate noncompulsory standards to which many nations will voluntarily subscribe. The Food and Agricultural Organization (FAO), International Labor Organization (ILO), World Health Organization (WHO), and International Maritime

Organization (IMO) among them have much authority that coordinates and encourages without being "compulsory" in a sense familiar to domestic legal systems.[3]

THE IMPEDIMENTS TO TREATY FORMATION

The question to turn to now is this: Considering the worldwide concern over the environment, and the relative advantages of treaties, why are there not more worldwide agreements? What are the impediments that led to such disappointments at Rio?

There are several reasons why environmental treaties—truly effective treaties—are hard to negotiate. Some of the difficulties stem from problems that hinder the formation of any sort of international accords in any subject area. Others are specific to particular areas of current environmental concern. Let us look at these two sorts of problems in turn.

Some General Problems in Treaty Formation

CONSTRUCTING COALITIONS

The first general impediment to treaty formation arises from the freedom nations retain not to be bound. Within an ordinary democratic society, a decision by the majority binds all members, including dissenters. But under international law, states are bound only by the rules to which they have expressly consented, with a few exceptions for the most basic precepts of the world community. For example, the obligation of nations to honor their commitments and the immunity of diplomats from prosecution by the host state are considered so basic that each state, on achieving statehood, can be presumed to consent to be bound by them. But rules of this sort are largely irrelevant for the contemporary environmental situation, since no one pretends that the principles of modern conventions on weapons testing or sulfur dioxide are part of a long-established body of law that has become universally applicable.

These far-ranging prerogatives of sovereignty exacerbate the "free rider" problem, a strategic obstacle inherent in any effort to arrange for the supply of a public good (public safety, protection from disease, parks, and so on). The benefits of such goods being indivisible (I get as much police or ozone-shield protection as anyone else, whether I chip in and pay for it or you do), it is a principle challenge of all governments to arrange for their supply. Especially as societies are large and diversified (and what society is larger or more diversified than the world's?), each benefiter is tempted to withhold its share of support in the hope that the other benefiters will make up the deficit. Majority rule and taxes provide a way of overcoming the problem—-where they prevail.[4] But in the world community, not majority rule but a sort of unanimity rule (no one goes along unless they choose to) prevails. There is, indeed, no "legislative" body, and in all events no system of taxation, as such, exists. Hence, cooperation for the overall public good is all the more difficult to achieve.

As with any public good, these global ones confront regulators with two major issues. What is the right level of supply (which can be viewed as the optimal level of fishing, emissions, monitoring, etc.)? And, how is the cost of achieving that right supply to be apportioned among the benefiters? Any answer, particularly one hardened into effective regulation, will require widespread international cooperation. But it is the achievement of that cooperation which free-riding and other strategic behavior frustrates. In so large and diverse a community, each member is tempted to conceal preferences (understate its true interest in a proposal) and withhold a fair share of support, in the hope that the other benefiters will shoulder the burden and make up the deficit.

These aspects of the world community—150 sovereign states and growing, all politically equal—are a general impediment to a cooperation. But efforts to secure cooperation are hardly futile. In some established forums, such as the U.N. General Assembly and within certain international agencies, majority rule or variations upon it are built into the practice.

Moreover, in many contexts a convention does not need universal support in order to be effective. Although 117 states have now signed the 1963 Treaty Banning Nuclear Weapons Tests in the Atmosphere, in Outer Space and Under Water, considerable effectiveness was virtually secured with the agreement between only two powers, the United States and the Soviet Union. Coal burning, a principal source of greenhouse-creating CO_2, presents a similar opportunity. In view of the fact that nearly 60 percent of the world's coal is consumed by only three nations—China, the United States, and the former Soviet Union (in that order)—an agreement among those states on coal alone would have a noticeable impact on the global atmosphere.* And, as we have seen, many problems, such as pollution of the Mediterranean, can best be established on the regional, not global level. Thus, the first point is: not every nation has to sign an international convention for it to make a substantial difference.

THE QUESTION OF COMPLIANCE

A second general problem is how to get nations that assent to a treaty to comply with their commitments. As noted above, there is no sure way to compel a nation to appear before a national or international tribunal to answer charges against it. And even when nations do appear to answer charges, the sanctions that international law makes available tend to be far less intimidating than those available to the ordinary nation internally. Save, perhaps, for the case of war crimes, the international community does not itself mete out direct punishments to individuals (such as imprisonment) or impose punitive damages or fines on the nation. There is not even what would pass, in a domestic court, for an injunction.[5] Even damage awards against states, while available, are infrequent, slow to extract, and generally, by U.S. standards, leanly calculated when awarded. What is there then to make nations comply?

Compliance with international law, in the absence of any international sheriff with badge and pistol, is a large subject on which a great deal has been written. (The eminent English philosopher John

---

* As we intimated in chapter 4, there is little likelihood that the coal-burning states will be inclined to sign such a treaty, unless, at a minimum, there are linked reductions by other nations of other greenhouse-gas-affecting activities.

Austin declared that, lacking these elements, international law should not be honored as *law*.)[6] But, as with sovereignty and the unanimity rule, the limitations ought not to be overstated.

Granted, the compliance problem is serious. Any nation that refrains from joining a ban on trade in ivory or rhinoceros horns stands to reap the benefit of the higher prices that the reduced supply will invite. Nations that refuse to participate in a convention restricting fishing have that many more fish to catch. The prospect of these "rewards" for noncooperators undermines efforts to form an effective coalition.

But the foundations of the world order are not as flimsy as they are often depicted. Even the commonplace that there is no international criminal liability must be understood with qualification. There may be no international criminal court, but international agreements can oblige the parties to use their own judicial systems to punish violators. For example, the 1954 International Convention for the Prevention of Pollution of the Seas by Oil obliges the offender's state to provide penalties "adequate in severity to discourage any such unlawful discharge and [which] shall not be less than the penalties which may be imposed under the law of that territory in respect of the same infringements within [its] territorial sea."[7] Then, too, while international law traditionally depended upon nation states to press enforcement, compliance, especially in the United States, is increasingly achieved through suits by nongovernmental organizations and private parties.*

Moreover, sheriff or no sheriff, there is no reason why those nations that do sign an international convention have to sit idly by and let the nonsigners "free-ride" on their efforts. The international com-

---

* This factor tosses a curious complication into treaty-making. No nation enforces all the laws on its books. Some nations are prepared to sign treaties obligating them to enact stiff laws, but without any strong commitment actually to enforce them. The U.S. tends to enforce its environmental laws far more than most; and, indeed, if the U.S. government fails to do so, there are liberal rules empowering non-government litigants to step in and either sue themselves or force the government's hand. This may make the U.S. hesitate to sign agreements that other nations will put their signatures to more cavalierly.

munity has its own special carrots, such as technology transfers and economic assistance. And, that failing, the world community has its own sticks. In September 1992, in the wake of outrage over aggression and atrocities in the former Yugoslavia, the United Nations expelled the Serbian-dominated Yugoslav federation. In response to apartheid, South Africa has been suspended. Less draconian measures range from censure by the United Nations or other formal body, to making a cheater ineligible to receive loans from multinational development banks. There is even the potential "sanction of nonparticipation," that is, the exclusion of delinquent states from participating in international scientific and technical organizations and programs.[8]

In addition, the noncooperator's products can be subjected to special tariffs contrived to offset its unfair advantage. At least one bill introduced in the Senate would put the United States in a position to impose countervailing duties on just such a basis.[9] But the concerted boycott involving a larger community of nations is the more common rejoinder. Countries that sign the Convention on International Trade in Endangered Species (CITES) agree to prohibit trade in endangered species and their products with nonsignatory nations, thereby restricting (unfortunately, not eliminating) markets. The same sort of nontrade provision exists under the Montreal Ozone Protocol. The agreement to reduce CFC production would be undermined if signatories were allowed to sell production technology to nonsigning nations. But the Protocol closes this loophole by prohibiting the signatories from engaging in such trade.[10]

The informal constraints of good neighborliness have a definite influence, too. Although nations are under no compulsion to go along with the majority, they generally wish to be considered responsible members of the world community. Indeed, on one theory of hegemonic power, the leading nations may be particularly sensitive to do the right thing, even if it entails, as it often will, bearing a disproportionate share of the financial costs in exchange for the glory of world leadership.

The desire to be a good—or leading—world citizen is not only a motivation to sign international accords, it may even induce some nonsigning nations to meet a convention's standards even though they have not formally bound themselves to do so. Both the United States and the United Kingdom have declined to become formal parties to the 1985 Helsinki Sulfur Protocol, under which states commit to achieve 30 percent reductions in transboundary fluxes. Yet both countries, mindful of world opinion, are careful to stress that they have met the requirements with substantially reduced sulfur dioxide (SO_2) emissions in the 1975–84 period, the United States by 33 percent and the United Kingdom by over 40 percent.* Community pressures have operated in the same way in the area of atmospheric nuclear weapons testing. Neither France nor China has signed the 1963 test ban treaty but both have displayed some sensitivity to its demands.

Of course, even in the face of these legal levers and "social" pressures, nations may be willing to suffer a little tarnish in image when the benefits of lawbreaking, in terms of some national self interest, are viewed as worth it. (Just as ordinary people sometimes calculate an advantage, on balance, in breaking a lease or contract.) But the dominant fact remains: even in the absence of many familiar formal sanctions, most nations are inclined to advance global interests and to honor their undertakings, if only because there are reciprocal advantages of being considered a reliable citizen of the world community.

Yet undeniably, these two general problems—getting nations to enter into conventions, and then getting the signers to comply—

---

* The U.S. remains a nonsignatory, even though the Clean Air Act of 1990 calls for a 50 percent reduction in sulfur dioxide emissions from 1991 levels by 2000, in addition to stricter health standards designed to cut smog in urban areas. One reason why a nation might resist formally accepting a treaty, even while intending to observe its present terms, is concern that if it should enter a formal convention, it may be caught in a web of increasingly onerous regulations over time. Britain's minister of state at the Department of the Environment, the Honorable William Waldegrave, was undoubtedly sincere in claiming that "even though we were not able to sign the protocol, we believe that our actual performance will be equal to—or better than—that of a number of countries who did." See "The British Approach," *Environmental Law & Policy* 15 (1985): 112.

have two important consequences on the shape and mood of international law.

The first is that treaty targets tend to be set at relatively undemanding levels: making ambitions "realistic" is the easiest way to muster consensus and assure compliance. The 30 percent reduction in SO_2 agreed to at Helsinki represented not the level that would check acidification (some environmentalists believed that would have required a 90 percent reduction) but simply a figure that the governments involved could live with.[11] Indeed, SO_2 emissions had been decreasing even without a treaty, at about the same pace the treaty confirmed.[12] Second, as we have seen, the sanctions for failure to comply, if any, tend to be more vague and more mild than their domestic counterparts.

Special Obstacles Confronting Environmental Agreements

If we go beyond the general problems that beset the making of all international conventions and focus on the obstacles to the negotiation of contemporary environmental treaties, we form a better idea of why the UNCED Earth Summit at Rio fell so short of its advance billing. The fact is, the major global environmental problems that drew the world leaders to Rio—loss of biodiversity, thinning of the ozone layer, thickening of the greenhouse blanket, the pressures on living marine resources—each contain the germ of potentially divisive conflicts. There are reasons why nations may decline to enter global agreements on them in the most earnest good faith—that is, even putting aside displays of strategic behavior that one has to anticipate in any multiparty negotiation.

JUDGMENT ABOUT IMPACTS

In many negotiations over the environment, there can be, as we saw in Rio, perfectly good-faith disagreement as to whether some alleged phenomenon is actually taking place on a serious scale—for example, whether we are actually witnessing a statistically significant increase in world temperature, or a decline in total marine resources. The uncertainty is amplified when future effects are projected:

When, if ever, will the level of atmospheric CO_2 double, and what consequences would such a doubling have, for, say, soil moisture and sea level? When so many scientists are in disagreement, it is hard to expect diplomats to put their fullest weight behind ameliorative programs that compete for resources with other legitimate, and in most ways more immediate, demands—such as the elimination of AIDS, poverty, drugs, and so on.

JUDGMENT ABOUT CAUSES

Even if the global community is in general agreement that a certain perceived effect (a thinning of the ozone shield) is truly occurring, rather than apparent or illusory, negotiation may be hampered by disputes over cause.

For example, periodically the oceans are marked by "blooms"— sudden swellings in oceanic algae populations. These episodes have lethal effects on marine life (and occasionally, indirectly, on people who consume them), either through depriving other marine life of adequate oxygen or, in the case of "red tides," poisoning them through associated toxins. There is considerable concern that the algae blooms are fostered by man-made pollution, including runoff from fertilized agricultural land, sewage discharge, and possibly acid rain, all of which can inject surges of nutrients that swell the algae population.[13] On the other hand, "red tides" (as they are often called) have been reported periodically through human history. Exodus 7:21 records an episode as one of the plagues visited on the Egyptians—long before our tampering with the environment had reached significant levels.[14]

Moreover, even if human activities, as seems likely, are exacerbating the algae blooms,[15] there is considerable room for uncertainty as to what chemical agents and which nations' activities are causally responsible. Nation A may point the finger at nitrogen emanating from Nation B's agriculture; Nation B blames phosphorous and trace metals from Nation A's industrial waste—and there may well be a grain of truth to each position.

Similarly, while human activity may be affecting climate in significant ways, the principal determinants of global climate remain natu-

105

ral forces, as we saw—most notably solar radiation, the tilt of the earth's axis with respect to the ecliptic, and the eccentricity of the earth's orbit. These relations continually shift over long cycles, signaling ice ages and their retreats.[16] There are enormous and wide-ranging climatic effects of such other natural processes as solar activity, volcanic debris,[17] and the El Niño, as well. All these "background" processes complicate efforts to determine the extent to which any perceived changes in the environment can be pinned on human activity—and therefore be subject to mitigation by changes in human conduct, through law.

NEGOTIATING LINKAGES

Even where the contribution of human activity has been segregated out, negotiation will often face considerable controversy over questions of linkage. No one doubts that the blanket of greenhouse gases is thickening. But the blanket consists of a number of gases. Carbon dioxide, largely a by-product of fossil fuel burning, is the most abundant and has therefore attracted the most attention. But methane, substantially connected to agriculture, has been increasing more rapidly, at least until quite recently.[18] While the expected residence in the atmosphere of each carbon dioxide molecule may be ten or fifteens times that of a methane molecule, each methane molecule is perhaps fifty times more effective a "blanketer" of outbound energy, molecule per molecule, than CO_2.

This background complicates climate negotiations considerably. The U.S. negotiators, and perhaps other major fossil fuel burners (China, Russia, and Germany), have understandable reservations about diminishing their fossil fuel use unless these reductions are linked to sacrifices by those who affect the greenhouse blanket in other ways. The others would include India, the primary producer of methane, and Brazil, whose massive forest burnings not only result in major CO_2 emissions, but obliterate one of the planet's major "sinks" for drawing excess CO_2 out of the atmosphere. These demands for linkage do not appear to be unreasonable, considering, for example, the suspicion that at the present margin a reduction in methane could be achieved more cheaply.[19]

The various greenhouse gases (GHGs) can be related to a common blocking power index on the basis of their expected atmospheric residences, absorptive powers, and various yet uncertain climatic feedbacks.[20] Several methodologies for doing so are under consideration. But to give a flavor of the feelings that the negotiators will have to face if they insist on pressing linkage, before the Rio conference an Indian policy center decried as "greenhouse imperialism" a study by the World Resources Institute that dared nothing more than to compare the emissions of the less developed countries (largely methane) with the emissions of the industrialized world (largely CO_2). That, the center complained, was to compare "survival emissions" with "luxury emissions" and would be "criminal" if used as a basis for international action.[21]

NEGOTIATING BASELINES

The solution to many problems requires agreement on some baseline upon which negotiated restrictions will operate. To sort out the alternatives is to enter a diplomatic and moral quagmire. Take GHGs, once more. Should a reduction formula be proportioned to historic levels of emission, laissez-faire levels (that is, projected levels absent agreement), Gross Domestic Product (GDP), territory, population, a threat point (that is, what each nation *could* emit if it, say, exploited its coal reserves at full pace without scrubbing), or someone's calculations of need, that is, "survival" versus "luxury" emissions?

Temporal baselines are every bit as knotty. Every negotiation that looks to an agreed-upon rollback of emissions has to agree on how far back to go to find the benchmark year. Ordering cutbacks from levels current at the time of negotiations may convey a perverse message: it suggests to any nation considering voluntary reductions that it may be required to make even deeper, more expensive cuts (from its new, self-imposed baseline) later.[22] Frictions are rife. One of the draft agreements in the nitrogen oxide (NO_x) negotiations called for nations to freeze emissions to the 1987 levels by 1994. But the United States balked, pointing out that it had already cut NO_x emissions substantially under its Clean Air legislation, and taking the position—ultimately successful—that each nation should be al-

107

lowed to pick a base year, presumably the year in which its output was at its peak. Great Britain raised a comparable objection in refusing to join the sulfur protocol.

COSTS

To get nations to cooperate, there has to be some consensus that the benefits of the undertaking merit the costs. There are several reasons why the costs may not seem "worth it" to key negotiators, with consequent pressure on the consensus.

Most obviously, a nation may object to some proposal on the grounds that it is being asked to bear a disproportionate sacrifice. At Rio, the United States, as the heaviest producer of CO_2, put up the largest resistance to CO_2 constraints. On some estimates, to restrict carbon dioxide emissions over the next century to 1990 levels (one of the Rio proposals) would have cost the United States in the range of $800 billion to $3.6 trillion.[23] That is the principal reason why the United States, while assenting to the vaguely aspirational Framework Convention on Climate Change (which the Senate approved by voice vote in October 1992), used its influence to squeeze out any firm reduction targets. On the other hand, the United States was second to none in the fight against tropical deforestation; the leading deforesters, such as Brazil, the Philippines, and Malaysia, foiled in their bid to plug the United States' carbon, succeeded in derailing the antilogging measures.

THE UNEVENNESS OF IMPACTS

Even among the hazards we refer to as "global"—say, ozone depletion and climate change—the damages are projected to fall unevenly across regions. As a result, some nations will place a higher value on collaborative action than the others.

The world witnessed some such regional divisiveness at the London ozone negotiations. The thinning of the ozone shield poses its gravest threat to light-skinned populations living at high latitudes; that is because the thinning is most pronounced toward the poles and its principal direct human health hazard is skin cancer of a sort to which fair-skinned people are most susceptible.[24]

This is not to imply that dark-skinned midlatitude populations, among which sun-induced skin cancer is rare, positively *want* an ozone thinning. But the phasing out of CFCs poses trade-offs in terms of development, and it was not surprising that at the London negotiations it was the Scandinavian countries, Germany and Canada in the Northern Hemisphere, and New Zealand and Australia in the Southern, that appeared to take the lead in pressing for the most rapid phaseout of ozone-depleting agents.

When we turn to climate change, the situation is even more complex. After all, everyone agrees that ozone thinning is bad; people simply differ on the sacrifice warranted to achieve any particular level of constraint. By contrast, some nations may actually believe (or can credibly claim) that the odds favor their benefiting from greenhouse warming. This can be expected to hinder negotiations even more severely.

To illustrate, recall that if the earth heats up, precipitation worldwide is expected to increase.[25] A warmer, moister, more carbon dioxide-rich atmosphere, together with longer (frost-free) growing seasons, is generally viewed as favorable to the growth of biomass overall (with some variable impacts on plants depending on species and local conditions).

Consider what the position of former Soviet republics is likely to be. According to a study by the United Nations Environmental Programme (UNEP), the effect of a 1.5° C increase in temperature in the Central European area of the former Soviet Union would be "a 30 percent increase in wheat yield. Additionally, the area suitable for wheat cultivation would increase by 26 percent, providing an overall increase in wheat production of 64 percent."[26]

The old Soviet region stood second among nations in contributions to global CO_2 additions[27]—and therefore might expect to bear one of the highest burdens in any reduction agreement. Can we expect nations so situated, with all their other worries, to pay more for heat and energy today in order to have less to eat tomorrow?

Indeed, on the basis of present knowledge, can we really dismiss the possibility that carbon consumption, moderately masked by available technology, could not meet Kaldor-Hicks criteria for efficiency (see chapter 6)? In ordinary parlance, that means while some will win and some will lose from continued GHG emissions, it is an open question whether the gains from enhanced agriculture will not dominate the losses from floods, etc., both across space and across generations. The very fact that it is an open question hinders—and indeed, ought to hinder—the garnering of a worldwide consensus that radically stringent carbon-constraining policies are an *ideal*— even before we get to the difficulties of implementation.

DISCREPANCIES IN VALUES

Across the world, one hears strong support for the environment. But the real question in negotiation is how sturdily environmental values, such as respect for species and lifeless tundra, will be maintained when they come in conflict with the value of human lives and qualities of living. The values that are put in play vary enormously among individuals within our own nation, and the range varies all the more as we move across cultures. Some cultures regard the earth and life upon it as relatively insignificant in comparison with eternal paradise; others regard all of Nature (of which *homo sapiens* are just another ingredient) as sacred; still others are relatively homocentric. Some nations value whales in the ocean; others, on their plates.

Such valuative differences are compounded when long-term effects are accounted for. Many of today's controversial actions, such as the storage of nuclear wastes and the elimination of wilderness areas, involve a shifting of risks onto the unborn. This raises a series of terribly complex problems, on which different nations are apt to differ. There are of course the straightforward empirical uncertainties. We cannot be sure what the remote effects of our actions will be, particularly in light of the technology that will be available to our distant progeny. Will they really care that the oceans are whaleless? They may be far better off than we, looking back on our lives (as we regard those of the late Middle Ages) as short, nasty, brutish, and deprived. And even if we do conclude that some of our present actions will affect our descendants in adverse ways, how do we mediate between their interests and ours? We may be leaving them hazardous waste sites; but we are also bestowing on them Shakespeare and Beethoven and VCRs. ("What," some ask, "have future generations done for us?") These raise knotty inquiries for philosophers, on which little consensus can be expected worldwide.

Finally, there are different valuative postures that arise not so much from differences of cultural and philosophical attitudes as from sheer differences in wealth. Those under the pressures of poverty (and international debt repayments) are destined to bring a sense of urgency to the present and to discount the future correspondingly. Of the populous nations, Egypt and Bangladesh probably have the most reason to worry about sea-level rise in the next century. But they also face terribly pressing economic problems right now. If each were given a "budget" and told that they could apply it either to reduce the risks to future generations or relieve the present suffering, who can really doubt which they would choose?

STRATEGIC OPTIONS

Suppose that nations can agree on all of the above: that some effect is taking place; that some identified agent is the cause; that the consequences are bad; that the benefits of eliminating the risks warrant a joint global effort, even if the costs rise to such and such an amount; and so on. Negotiations can still break down over strategy—on how best to tackle the problem.

Some international accords prohibit certain conduct outright—for example, those that forbid trade in endangered species, or nuclear-weapons testing in the atmosphere, or war crimes. But note that prohibition is feasible in those instances because the forbidden activity is almost universally condemned. There is no redemptive value in mistreating POWs. By contrast, much of the global degradation results from activities that, while hazarding the environment, have associated benefits. Transporting oil by tanker and burning fossil fuels are representative. The underlying activities cannot be eliminated outright without significant sacrifice in global welfare. In these "gray" areas, the question is not to ban, but how to regulate the activity so that competing legitimate interests are maximized. In this area, nego-

111

tiators have to weigh the merits of economically more sensitive control techniques, including user charges, marketable pollution rights, and so forth (see chapter 7).

The technique favored—for example, whether to ban ocean dumping or allow it subject to permits, or to empower the injured to sue—will of course reflect competing national priorities regarding economic growth. But the fashioning of relief can also be impeded by unsettled factual questions. Consider ocean dumping. While the loudest voices call for a halt to the practice, there are respectable voices, including the National Research Council of the National Academy of Sciences and the National Advisory Committee on Oceans and Atmosphere, who reason that deep-ocean dumpsites, carefully monitored, may be the best resting ground for some of our worst waste.[28] The issues are highly complex, but their point, essentially, is that while fail-safe containment or destruction would be ideal, it is presently unachievable; deposited on land, the waste is ordinarily more of a hazard to human populations; and that, given the law of gravity, much of our waste is now reaching the deep ocean anyway, meandering through river and wetlands, doing more damage on its way than if we just deposited it in the deep oceans at the start.[*]

Strategic debates are shaping up in regard to biodiversity as well. Mathematicians have questioned present emphasis on preserving the most endangered species: they maintain that biological diversity should be measured by the evolutionary distance between available species, and what conservationists should aim to optimize is a function of that distance.[29] A prominent economist has reminded us that efforts to save the most highly endangered species are apt to be the most costly and futile; when the costs and benefits are considered, rational policy should include making presently unendangered spe-

---

[*] This argument is persuasive only if we regard the quantity of waste as fixed. But it need not be. Free and easy deep-sea dumping fails to encourage the development of alternative techniques for waste management and results in more deposits than if there were a charge. A system of charging ocean dumpers for the privilege and the application of the funds raised to repair the global commons areas are discussed in chapter 9.

cies safer.[30] And a zoologist has queried whether the single-minded focus on rain forests may not be misdirected: worldwide, dry areas are more threatened and are more important custodians of "higher taxonomic categories"—notably mammals, which store, species for species, more genetic information than the lower life with which the wetter jungles teems.[31]

Strategic agreement can be plagued, too, by the existence of diverse cultural and national "tastes" toward various specific strategies. We would have to defer to the cultural anthropologists to opine the depth and breadth of their influence. My impression is that the United States displays a relatively high preference for clear rules spelled out in advance, is accustomed to settle differences through litigation, but is prepared to experiment with economic instruments such as pollution rights. Other countries are more accustomed to living with cloudier guidelines (which means more administrative discretion), inclined to mediation over litigation—and regard economic instruments as the cryptic handiwork of the capitalist devil.

THE NORTH-SOUTH CONFLICTS

As we saw at Rio, the most fractious dissensions arise from North-South conflicts. The nations of the earth enter into global negotiations with a profoundly unequal distribution of wealth. As a result, two sets of issues are in play—each of which would be intimidating even if presented singly. First, there are the public-goods questions: how to define and provide for the "right" rate of GHG emissions or ocean fishing or habitat preservation. In that dimension, the negotiations assume the character of a cooperative, potentially positive sum game. What is sought is an agreement that improves upon non-agreement, with the benefits and burdens of mutual cooperation so distributed that no nation emerges worse off than when negotiations began.[32]

Note that on this first view of the process, the negotiators are regarded as accepting the prevailing wealth and rules as a given—the "no agreement point" that serves as a baseline from which advances are sought. Any questioning of the starting point is severed for a focused discussion in the context of agencies specifically charged

113

with development and redistribution, including the development banks and various U.N. agencies.

For many of the LDCs, however, the starting point is the question. For them, the remote vulnerability to an environmental collapse, while worrisome, is in general secondary to the immediate threats of stagnation and poverty. The LDCs are thus not inclined to delink the issues. Rather, negotiations over climate or anything else are seen as an opportunity to reduce the underlying disparities in wealth and power.[33] Some nations, the rich, *should* emerge worse off, having transferred some of their affluence to the poor.

It is this tension between rich and poor that marked the exchanges at Rio. Those focusing on the environment (largely, from the environmental nongovernmental organizations [NGOs] and some First World representatives) wanted to maximize environmental protection—as much protection as available money could buy. The poor nations showed that they want, foremost, to pull themselves up from poverty. It is that conflict between the environment focusers and development focusers that gives rise to talk about "sustainable development," a term meant to suggest that conservation and development can go hand in glove. And, indeed, the environmental literature is full of case studies in massive and mindless developmental projects that managed to set back the local economy at the same time they were destroying the environment.[34] There undoubtedly exist *some* projects where developmental and environmental goals are fully congruent, and these should be identified and vigorously pursued. But more often there are tensions that cannot be obscured by diplomatic rhetoric. Environmental quality, economic development, and, for that matter, distributional justice are all potentially in conflict. In particular, the more a nation develops, some poverty-linked environmental problems, such as desertification, may be mitigated. But a larger number associated with carbon and chemicals—and these are the pollutants with the largest transfrontier effects—is positively linked. We may be able to lessen the correlation through technology transfer, for example, but we are not about to eliminate it.

Nor, however, can those tensions be ejected from North-South negotiations by dividing some talks into "the environment," to be attended to by environmental agencies such as UNEP, and others into "development," to be relegated to the jurisdiction of the International Monetary Fund (IMF), World Bank, and traditional developmental agencies. While the values of environment and development often conflict, they are also inseparable. Restrictions in the name of the environment, whether on fossil fuels or methane or anything else, are destined to affect rates of development and relative distribution of wealth.

The fact is, too, that the bargaining power of the Third World collapsed with the end of the Cold War. The developed nations are turning attention inward, and giving the old Eastern bloc an inside track on much of what is left over. The LDCs do not have many cards to play. Developmental assistance is provided on the developed world's terms. In environmental negotiations, by contrast, the developed countries are more anxious about the global environment than the LDCs whose cooperation they are courting. This gives the LDCs something to bargain over. And it is quite understandable they might want the bargaining to be opened up beyond the international public goods (the ideally thick greenhouse blanket, ozone shield, and level of biodiversity, whatever they may be), and to encompass some raw reallocation of wealth. If the poor get more assistance from the rich, they can make the funds available for local environmental challenges such as desertification and dirty water, as well as, of course, food, health care, and infrastructure. As in the Law of the Sea context, in which the same North-South clash arose, adding cards to the deck raises the risk that the environment will get lost in the shuffle.[35] And it is easy to show that in any negotiation in which majority rule prevails, a coalition of poor voters, provided with ample freedom to define the issues, probably *will* go beyond an "efficient" solution to the public-goods problem and wind up redistributing a little wealth in their favor.[36]

On the other hand, the insistence on linkage must appear to the LDCs no less legitimate than the United States' occasional gestures linking human rights to trade.[37] Adding pure redistribution to the agenda adds another unit of currency for bargaining, and hence another dimension along which advances in aggregate welfare may be sought—not a bad dimension, given plausible assumptions about the declining marginal utility of wealth. That is, if we admit the legitimacy of interpersonal comparisons of utility, it seems credible that a transfer of $1 billion from the United States to Bangladesh would bring about a net increase in welfare, their population gaining more in utility than ours loses.

In sum, even if an environmental convention provides an imperfect framework for resolving inequalities in global wealth, some degree of North-South economic conflicts cannot be avoided. The UNCED Earth Summit was convened, after all, in the name of environment *and* development. But the broadened focus at Rio proved to be—and will continue to be—an obstacle to achieving outcomes that are ideal *environmentally*, at least as environmentalists understand that term. From the environmental perspective, and probably from that of development, returning *development* policies and strategies as much as possible to the IMF, World Bank, and traditional assistance agencies would tidy both of the working areas considerably.

UNDER WHAT CIRCUMSTANCES ARE INTERNATIONAL NEGOTIATIONS LIKELY TO SUCCEED?

It is not my intention to add to the postmortems on the 1992 meeting in Rio. With thousands of government officials, NGO representatives, observers, activists, media, and movie stars gathered together in one tent for two weeks to repair every problem under the sun, it is small wonder that the Earth Summit disappointed the expectations that no one should have held for it. It is rather a surprise that agreement on anything survived collapse under the weight of so ample an agenda and glamorous a multitude. I have made clear enough my views why climate change may have been a poor—perhaps the worst possible—choice for a focal point.

The disappointments of Rio should not obscure the fact that, even in the face of all these potential conflicts, declarations are proclaimed and treaties do get signed. Thus, in closing this discussion of treaty formation, a better use of history—and horse sense—is to speculate why progress has been made in some areas and not in others. Specifically, if we are not to tackle every problem at once, what characteristics make an area of negotiation a propitious target for successful international negotiations?

Let me suggest (in some instances, review) a few features that appear to favor success of multilateral agreements—not necessarily if displayed singly, but certainly where they appear in combination.[38] The first is that the *agreement doesn't (yet) matter*. Several early conventions regarding outer space, the moon, and celestial bodies were hammered out successfully because negotiations began when their effects were still distant in space and time. This suggests the advisability of getting an early start—anticipating rather than reacting—before potential vetoing constituencies have an opportunity to form.

A related point is that a treaty is clearly the easier to conclude, *the fewer the number of nations whose cooperation is required to make it effective*. This was true of the 1963 Treaty Banning Testing of Nuclear Weapons in the Atmosphere* and also the CFC cutbacks under the Montreal Ozone Protocol: in both cases, many nations ultimately signed, but success was sealed with the assent of a relatively small core.

It appears, too, that treaties come easiest when *the activities to be regulated are on the global commons*. Regulation of pollution *on* the high seas has been achieved far ahead of controlling pollution *of* the high seas from land-based sources. Partly, this results from the world community's readiness to recognize the legitimacy of international restrictions on activities that affect, particularly those that take place on, the shared space of the global commons. But the failure to stem land-based pollution also reflects another, more obvious factor: agreements are more likely when *the costs of restrictions are low*. That, as we saw, was a major distinction between controlling ozone-depleting agents and controlling GHGs.

---

* Not only was the number of critical players small, the major powers had a mutual interest in protecting their nuclear monopolies.

There is a suggestion, too, that coalitions may be strengthened if *it is feasible at least to monitor (and if possible, to enforce) compliance*. Critical to the banning of nuclear weapons testing was the conviction that violations could be monitored. There was considerably more resistance to the banning of underground testing until is was shown that seismic equipment for measuring earthquakes could detect underground explosions as well. The ban then became effective just because violations could be detected, even though there were no "real" liabilities for violators. Monitoring difficulties are another factor why the Economic Council of Europe (ECE) had a hard time producing a transboundary NO_x agreement as a sequel to its SO_2 accord. The bulk of SO_2 is produced by a relatively small number of major coal-burning sources, such as utilities. NO_x derives from so many point sources—automobiles are a major contributor—that NO_x was not as easy to clamp down on.[39]

It helps, too, when *the consensus as to need for action is strong* and *the technical experts are in agreement*. Scientists and other technical experts do not make treaties; but consensus among experts certainly helps move negotiations along. This boost results not so much from any influence experts exert on the political process directly. It is more that the political path is straighter when the experts report, for example, that *technical fixes are not foreseeable*. To illustrate, the concern in the United States over sea level rise, and hence the willingness to negotiate a CO_2 agreement, would be stronger if there were no technical undercurrent that future engineers and construction technology will be able to defend such areas as New York City without insurmountable difficulty.

It may seem obvious, but it bears reminding, that *the proposed restrictions have to be practically achievable*. The CFC reduction levels reached at Montreal and London were made possible only when the prospect of CFC substitutes, principally HCFCs that at least at the time were believed to be generally acceptable, lifted the resistance of U.S. manufacturers.[40] The schedule for phasing out halons, used in fire extinguishers, is less ambitious, not because halons are less threatening to the ozone layer pound for pound, but because there is yet to be developed a comparable substitute—and fires are a continuing worldwide concern more lethal right now than ultraviolet radiation.

CONCLUSION

Where are we left by all this—the virtues of treaties, the various impediments to treaty formation, the circumstances that may be, hopefully, favorable for their growth? In some circumstances, as we shall see in the succeeding chapters, the obstacles are significant enough that they invite alternative approaches—specifically, unilateral action, and various styles of more modestly scaled bilateral and small numbers undertakings.

But the notion of more ambitious multilateral conventions will and should go forward. In that evolution, the impediments are best viewed not as absolute barriers to agreement, but as factors that shape the structure and timing and mode of international action.

For example, where conflict takes the form of disagreement over effects or causes, there can be a first-stage agreement to cooperate on modeling and data gathering. While nations can and do act separately in scientific matters, joint sponsorship of the efforts is likely to make the results more credible internationally and probably more accurate as well. While the facts are being gathered, treaty negotiations, even the execution of softly textured conventions such as "declarations of principle," can proceed.[41] But until the scientific uncertainties are resolved, progress toward stringent and widely acceptable restrictions is apt to be limited—and often rightly so.

On the other hand, in cases where the reservations arise from fears that "the other fellow's" cheating cannot be monitored, the answer is concerted efforts to improve the monitoring. In light of the difficulty of tracking causal links, such as whose chemical agents are showing up where, it is encouraging to consider recent advances in the ability of scientists to trace some particles and gases to their place of origin. In describing pollution of the Arctic, *National Geographic* reports that "the Arctic pollutants showed a 'signature' unknown to Western scientists. [One investigator found] arsenic, selenium, antimony, and indium, in a combination that pinned most of the pollution to a mineral-rich smelting region in the Soviet Union's Ural Mountains. Other investigators found dry-cleaning Freons and degreasing solvents used commonly by the Russians but rarely by Western nations."[42] Similar advances in data gathering and processing could facilitate global agreements appreciably, and should be fostered as a matter of deliberate policy.

Where the conflicts arise over evaluation and choice of strategy, the marginal value of gathering additional facts or improving detection is minimal. The arguments there are not over what is, but what ought to be. What is required, at the least, is a continuing frank dialogue on international values and community—a subject so large I reserve it for separate treatment in chapter 10.

Nonetheless, no one should doubt that even without "hard" sanctions backing them up, treaties, and even vague, aspirational declarations of principle, have significant effects on patterns of behavior in the international community. Indeed, no one should doubt the salutary effects in the mere process of bringing diplomats together to discuss global problems. One well-informed commentator, Charles Pearson, has observed of the 1972 United Nations-sponsored meeting at Stockholm that produced the Declaration on the Environment: "It is difficult to assess the actual results of Stockholm. In part this is because the process of conference preparation and participation can be viewed as an objective that was more important than the Declaration . . . itself. . . . Preparations for the conference . . . force[d] governments into awareness of international environmental issues."[43]

Twenty years from the Earth Summit in Rio—and twenty years is not long in geologic or diplomatic time—the same may be said. Awareness of, and even the appreciation of deepening conflicts has increased. This is less likely to mean that efforts to reach consensus will cease, as that the focus of efforts will shift. Important issues on which agreement is possible will be delinked, compartmentalized, and shunted to more favorable forums. There, details can be worked out in a more productive (and sometimes less public) negotiating environment. In many areas significant advances are less likely to be achieved by agreement on governing rules (even in the limp form of aspirational targets), as on the creation of an infrastructure of agencies, informed as much by scientists as by diplomats, that can oversee and guide, step by step, the development of particular regimes.

CHAPTER VI

The Economist's Prescriptions: Taxes and Tradable Permits

As WE HAVE SEEN, many of the maladies of the global environment are symptoms of frailties in the world political order. As long as the earth is partitioned politically, individual nations will be tempted to overexploit commons resources and dump their wastes somewhere on the other side of their frontiers. Thus, the international environmental effort is rightly concerned with reforming political relationships so as to reduce the aggravations caused by international conflicts.

But it is folly to suppose that if the nation-state system were to be replaced by a world government, the problems of the environment would vanish—or even necessarily become more manageable. If we examine modern governments, such as those in the United States, there is hardly reason for optimism that the larger and more distant the governing unit, the more responsive it is to its constituents, or the more successfully it resists capture by competing economically and culturally defined interests. Indeed, around the world, a lot of conventional long-established governments are ungluing from a less divisive concoction of religious and cultural factions than that which a true world government would have to contend with.

Yet, while a unified government with plenary powers to command obedience across the globe does not, and perhaps ought not, exist, it is helpful to come at our subject by imagining for a moment that there were such a One World Government (we will call it the Authority).* Then we can ask—to clarify our objectives—how would that

---

* One "provocative" (to put it mildly) illustration of a purely economic, one-globe perspective has already been recorded: the World Bank economist's memo that from the global point of view, it would be better if heavily polluting industries were encouraged to migrate from richer countries to poorer, inasmuch as the aggregate costs of pollution would decline. See note p. 68, supra.

Authority define the ideal state of the global environment that we in our less ideal world ought to be striving for? What conditions should the Authority aim for? What strategies would it be warranted to adopt?

THE ECONOMIC IDEAL

These are hard questions to answer. But one thing is clear at the start: the Authority would not ordain a pristine planet (how would it even define "pristine"?) if that entailed, as it certainly would, a total suspension, even reversal, of economic development. Nor, presumably, would its goal be to preserve each and every species, whatever the sacrifice. Fossil fuels would continue to be burned, forests logged, oceans fished, fertilizers spread. The real question is, how would the Authority decide how much of these activities to suppress?

Efficiency

The most constructive responses come from economics. The framework is to regard the world as composed of people with desires, resources with which those desires can be satisfied, and the technology and the techniques for combining them most productively.* Given a population and the stock of resources available to it—the pool of labor, the stock of forests, minerals, and other raw materials—not every desire can be satisfied. The goal is to do the best we can: to draw from the procurable base the most welfare-satisfying mix of outputs achievable. But how do we judge which of the possible combination of factors is better than any other? According to one

---

* I do not accept some of the implicit assumptions without qualification; as I shall discuss in chapter 10, the world is more than a playground for the satisfaction of *homo sapiens'* desires, nor is it correct to accept our desires as given, rather than capable of adjustment in the light of moral reasoning—such as reasoning about the place of humans in the larger scheme of things. But in this chapter I shall work with the economic assumptions which, although often disparaged by environmentalists, are enormously important conceptual and policy tools.

of the most widely accepted criteria, a combination is better—it is a *Pareto-improvement*—if it makes at least one person better off and no one worse off.* If, for example, a conservation measure such as weatherproofing of homes saves all participating homeowners money, preserves resources, and cleans the air, so that at least some members of the society are better off and no one is made worse off, the measure is efficient in the Pareto-improving sense. In plain parlance, in those circumstances, since no one is worse off, who is to complain?

Unfortunately, in a large, complex society, almost every action a government contemplates threatens to make at least *someone* worse off, and it is therefore in jeopardy of flunking the stringent Pareto criterion. A decision to preserve a forest ecosystem by banning logging will hurt, if no one else, the owners of the timber rights and the loggers. But even in that case, the Pareto criterion may still serve as a guide. We might insist that the logging ban provide (what the Fifth Amendment to the U.S. Constitution in some cases requires) that those who benefit from "taking" and conserving the forest, the community at large, compensate the timber owners for the loss of their property and the workers for their lost wages. If the benefit is large enough for the "winners" (the community at large) to pay the "losers" (the logging interests) so that, even after compensation, the winners still retain a positive gain from the move and the property owners and workers, duly compensated, emerge in no worse a position than they were in before, then the ruling transforming the logging area into a preserve is still efficient. If, conversely, the winners cannot reach into their gains, compensate the losers, and still come out ahead, the transformation is not Pareto-improving; but then, under those conditions, using normal economic criteria the society is better

* Named after the Italian economist, Wilfredo Pareto. States of affairs that can be subject to no further Pareto-improvements are deemed Pareto-efficient or Pareto-ideal. Ordinarily there is no unique Pareto-efficient point, but a whole frontier where none of the points can be improved on. In other words, once on the frontier, you cannot make anyone better off without making someone else worse off. Which point on the frontier is selected as "best" may be a matter of power or morals about which the Pareto criterion has nothing to contribute.

off with the property uses left as is. This, of course, is the theory bolstering the South's demands for compensation for preservation discussed in chapter 2: if the developed Northern nations want an LDC to preserve an ecosystem, let them pay for it. Both North and South can come out ahead.

As the community affected gets larger and more populous so that the gains and losses of official rulings and private activities are broadly and variably distributed, it becomes less and less feasible for the winners actually to identify and compensate the losers. Imagine trying to apply the Pareto criterion to the ozone protocols. Billions of people are affected across the world, some (including the not yet born) as winners and others (if only employees of affected firms and midlatitude populations that will pay more for refrigeration) as losers. Presumably the phasing out of CFCs offers net benefits. But it would be impermissibly expensive to determine who were the winners and who the losers, to measure their respective gains and losses, and to arrange for each of the winners to pay each of the losers. In other words, the costs of strictly observing the Pareto condition would eat up the gains, and *make* "inefficient" social decisions that are otherwise overwhelmingly advantageous.

One response is to have those realizing the major benefits compensate the major losers, if only approximately. That, in fact, is what evolved in the ozone negotiations. While individual losers are not paid off, the nations that are worse off for renouncing CFCs and switching to high-priced substitutes are in principle to be compensated by transfer of technology and payments from the ozone trust fund nourished by the principal benefiters, the white high-latitude rich (see chapter 5). Indeed, India demanded (but did not receive) over $2 billion as its price for signing the accord.[1]

But even such rough-justice side payments (or "bribes" as they are often infelicitously called) are not always practically possible. The choice is then between rejecting projects that promise significant net gains, or denying every single unlocatable loser its effective "veto power" (on the grounds that its position did not improve). The preferred theoretical response has been to relax slightly the definition of what is "efficient." Under the *Kaldor-Hicks* criterion, a move is designated efficient if there are net benefits, that is, if the gains to the winners exceed the losses to the losers, even if (on account of practically insurmountable transaction costs) the winners do not actually get around to compensating the losers.

To illustrate, using the Kaldor-Hicks criterion, a global Authority contemplating the phasing out of CFCs would ask whether the gains were sufficient so that the benefiters *could* compensate the losers (if they were able to get together), and still come out ahead.* If so, the world is probably a better place for having the accords, even though some people will be disadvantaged.

How do these criteria inform the critical questions of global pollution and other misuse of resources? We already observed that in the nation-state system, each country, calculating how much to spend on emissions controls, is apt to be altogether too frugal the more the detriment falls on some other nation downstream or downwind. And it is easy to sense why this might be so. But we can now look a little more deeply into the underlying structure of such situations as a prelude to evaluating some of the proposed policy responses.

Illustrating the Economic Analysis

Let us take, as an illustration, one of the great and most contention-breeding international waterways of the world, the Rhine. The Rhine winds 820 miles from the Swiss Alps to the North Sea, picking up a noxious brew of industrial effluents, chemical wastes, and agricultural runoff from Switzerland, Germany, France, and, at the bottom, the Netherlands.[2] To simplify the analysis, let us focus on only one of the chronic irritants that has arisen: chlorides ("salts"), largely from fertilizer production and use in Germany and France, are carried downstream at the rate of 1,200 tons an hour to the Nether-

---

* Note, too, that a move being Kaldor-Hicks-efficient is not destined to make it uncontroversially "good." There is room to object to the *distribution* of the net (uncompensated) gain; in particular, it is little comfort to a Third World country to learn that a particular convention that will set it back is, considering all the benefits to others including the First World, beneficial *on the whole*.

lands, where they corrode pipes in the local water system and afflict the Netherlands' famous tulip nurseries.

The immediate reaction of most people is to say that the Germans and French—the polluters—should clean up or pay the damages; after all, it is their potash manufacturers and users that are vexing the Dutch nurseries.

But a moment's reflection will show this commonsensical verdict may be a bit hasty. It is true that if the Germans were to eliminate potash production and the use of other fertilizers, the Dutch would be able to grow more tulips. But if the Germans make and use less fertilizer, they will not be able to produce as much grain. A German might ask, do the Dutch have more *right* to raise tulips than the Germans have to raise grain? Are the Dutch to be awarded the upper hand morally simply because they are saddled with the lower bank geographically? Agreed, the way the river flows, the potash happens to be in the ordinary sense "the cause" of the trouble—meaning that if the Germans put less of it in the river, there would be less mischief done downstream. On the other hand, if the downstream Dutch would only grow some crop less sensitive to salt than tulips, the harm would be less, too. Why isn't the Dutch decision to grow tulips equally the "cause"? Or should it turn on priorities: Was Germany salting the river before Holland was planting bulbs?

The economic analysis begins with the insight that these moral questions about cause and blame are not only difficult but, in a way, curious. The Germans are buying Dutch tulips and the Dutch are eating German grain. They—indeed, all five Rhine countries—are part of a larger community in which they all ought to be seeking opportunities for mutual benefit. A nice way to conceive the issue is this: if there were not five competing countries, but one (call it the European Community), what mix would it ordain of (1) tulips, (2) grain, and (3) river quality?

The path that the economic analysis leads along is this. The fact that the Dutch are suffering a dirty river is not bad because the river looks or smells foul, or because water systems and tulip production are impaired. Indeed, it may not be a "bad" outcome at all. It de-pends. If each German's desires are given equal dignity with each

Dutchman's, etc., then the first question is whether their joint resources are being used efficiently, or whether there is some way they might be recombined in a welfare-improving way. The river can be cleaned and tulips promoted—but it will be at a cost of higher farm-food prices. The persistence of salt in the river is bad *if* the river is dirtier (or tulip production is lower) than what the joint society would prefer, even accepting the higher costs. The "supply" of pollution (like the "supply" of a good we positively value, such as steel) should change according to the same criteria: Is it possible to rearrange the present "mix" of factors and outputs in a way that will make people better off? If so, then our present arrangement is "inefficient"—and for that reason wrong and should be changed.

Internalities and Externalities

But how do we know whether a particular mix is or is not efficient? The mix is suspect whenever there is a deviation between *private costs* and *social costs*. Presumably the German potash firms are paying their land, machinery, and labor costs, and therefore reimbursing the society the opportunity costs of those factors they are withdrawing from alternative uses. But they are not paying for the water quality they are "consuming," nor for the tulips. As a result, they are displaying lower costs on the firms' books, lower *private* costs, than otherwise. But that, we know, is because they are not facing up to—not *internalizing*—the full *social costs* their operations are imposing on the joint society. Because they avoid accounting for the value that society attaches to clean water and tulips, those factors are *externalities* to the firms (sometimes referred to as a "spillover effect" of the polluter's production). And since much of the detriment of the pollution is an externality to the German firms, the country has no incentives to force its firms to explore changes in crops and containment technology. On the contrary, under the circumstances, their firms are enjoying a competitive advantage over their rivals in any country that does try to internalize environmental costs on its farm industry.

Obviously, the larger society of which these countries are a part may suffer. But how would the Authority judge at what point it was warranted to intercede? We have now glimpsed the concepts from which the answer can be derived. At any point of environmental cleanliness (at any margin), there are some benefiters from further efforts to clean up—and some losers. *Movement in the direction of further cleaning the environment—removing an additional unit of pollution—is efficient, so long as the marginal social benefits from an additional unit of cleaning up exceed the marginal social costs of doing so.* Or, as we might have said above: so long as the benefits to the benefiters are large enough that they could compensate the losers and still come out ahead. Beyond that point, any further efforts to clean up—say, a million dollars to remove the last kilogram of chlorides—costs the society more than the gains are worth.

INSTITUTIONALIZING THE ECONOMIC IDEAL

The "Polluter Pays" Principle in Its Earliest Appearances

How to reveal and implement the economic ideal? At an early stage, the law began to drift toward what is today widely known as the "polluter pays" principle. Not surprisingly, application began only against the most blatant forms of damage to private property: courts took the notion of trespass, which had originated as an action to recover for *personal* invasions of another's property, and adapted it to award damages against those responsible for propelling debris, such as rocks, onto their neighbor's land, even if unattended by any bodily crossing of the boundary line.

Trespass never developed as a remedy for more subtle environmental "invasions," such as the discomfiture caused by odors or bright lights. The gap was taken up by the development of nuisance law, under which landowners whose properties were beset by loud noises or dust from a neighboring plant might get an injunction or damages against the offending activity if it interfered with the peaceful enjoyment of their property. If the neighbors won, the plant could resume manufacture only if it were prepared either to abate the nuisance or to pay the neighbors not to enforce their newly won rights, whichever was cheaper. (Of course, the neighbors would not be expected to "sell" their rights under the injunction unless the polluter's offer was at least equivalent to the damages the homeowners were anticipating.) If the company could not abate the damages or meet the neighbors' payment demands, it would have to shut down. And that might be taken as evidence, as a signal, that it should not stay in business, that the peaceful residential area was more highly valued. On the other hand, the polluting company—such as the German potash firm—might discover that there was such a heavy demand for its product that it could afford to pay for the damages it was causing the tulip farmers (or perhaps make some damage-reducing investments in its plants) and still operate profitably. If so, then there was presumptive evidence that the social benefit of its production exceeded the social costs it was inflicting. We have already seen several reasons why this sort of court-driven relief is flawed, particularly in the international context. In practice, there are not merely problems of proof, but also enormous complications that stem from the patchwork of court systems. In fact, among the suits spawned in the Rhine, in 1974 several Dutch horticultural firms instituted proceedings in Rotterdam for damages or an injunction against French potash mines allegedly responsible for dumping some 11,000 tons of unpurified waste salt into the Rhine every day, about 40 percent of the river's total salt content at the time. In separate litigation, two Dutch water supply companies sued the mines in French courts to recover $5 to $10 million to renovate and repair their corroded water pipes. The cases bounced around from court to court for fifteen years, producing volumes of testament to the frustrations of such litigation. They are not over yet.*

But aside from the drawbacks that international law aggravates, there is a deeper problem with this sort of antipolluter litigation. The

* For example, after the water companies won a measly $340,000, the appeals court in Colmar, France, reversed on the grounds that a causal relationship between the discharges in France and the damage in the Netherlands had not been proved. The court also rejected the water companies' demand that the damages in Holland be estimated by an expert and the case be moved to a higher court. See "French Court Rejects Tribunal's Award of $340,000 to Two Dutch Water Companies," *International Environment Reporter* (BNA) 11: 652.

polluter-pays principle in this form—internalizing the damages on the activity causing the harm—represents a rough *advance* toward efficiency; but it comes up short of the ideal.

Two problems have already been indicated. The first involves impairment of collective goods. Even if the polluter had to compensate all the individual victims who could prove legally recognized damages, the polluter would not be confronted with undistributed injuries to the public at large; these public spillover effects include the loss of a clean, scenic, fish-stocked river, and so on. Under the private torts system, the polluter dodges them.

The second group of problems comes up even in regard to the victims who can sue. If the law assures the victim that the polluter will pay for injuries (particularly on a "strict liability" basis) the victim has no incentive to take defensive measures. To illustrate the drawback, imagine that the least expensive way to halt salting of the river is to store the wastes on land, which would cost the upriver miners $2 million; but that for only $500,000 the downstream growers could treat the water to a point where it did tulips no harm, or perhaps switch to less salt-sensitive crops. From the whole community's standpoint, the least-cost solution is for the tulip growers, the "victims", to self-protect for $500,000. But the polluter-pays principle provides no assurance that this social optimal will be achieved. The growers may choose to run up the bill, relying on the legal obligation of the polluters to pay them whatever their losses.*

---

* The Coase Theorem (named after Professor Ronald Coase) suggests that as long as the liability rule is clear and the transactions costs are negligible, the parties—the miners and the growers—will bargain toward the efficient solution, no matter which party the law imposes liability upon in the first instance. That is, should the law impose strict liability on the polluters, they can be expected to pay the victims to take the efficient solution (to treat the water or switch crops downstream) if those responses are less costly than the prospective legal damages the polluters would otherwise face. But, conversely, if the law says the polluters are not liable, the growers will travel upstream to initiate negotiations to pay the polluters to take steps to abate the pollution; in that case, too, they will work out an efficient solution. However, in the global community, transaction costs—bringing the parties together—are hardly negligible, and whether the parties will be able to negotiate the optimal level of care in many cases is highly problematical.

In fact, the case for the "polluter pays" principle can be stood on its head. Two noted economists, William J. Baumol and Wallace E. Oates, point out that (1) if we take as the starting point of negotiation the status quo (the established patterns of pollution and other social activities), and (2) confine solutions to the set of pareto-improving moves, then for the polluter to pay the victim has got it exactly wrong: the victim should pay the polluter, if neither party is to be made worse off.[3]

Of course, in some circumstances, as where the polluter has acted with wanton disregard for the health and safety of others, we are right to make the wrongdoer compensate its victims, and even pay criminal penalties. Moreover, the Baumol-Oates point seems more telling in settling particular isolated disputes that have already arisen, rather than in legislating a rule that will govern a broad class of future situations. Nonetheless, in a whole range of more ordinary, long-standing polluting activities, the "victim pays" point deserves consideration. Indeed (here is the punch line), in the Rhine chloride negotiations we have been using as an illustration, it turned out that the least-cost solution was to store the salts on land. Therefore France and Germany stored, but the Dutch, although in the commonsensical view more sinned against than sinning, footed the largest share of the costs. And this, even though the solution directly contradicts the "polluter pays" principle espoused by the Organization for Economic Co-operations and Development (OECD).[4]

Pro Rata Cutbacks

Among diplomats, the favored strategy is to make each nation scale back from its present level by a fixed percentage applicable to all signatories. The pro rata cutback has the appeal of simplicity: the negotiators determine that total emissions are to be reduced by some overall target, say, 25 percent, and then order each of the nations to cut their emissions proportionately. Each national government, in turn, is expected to develop its own internal plan for achieving the national goal. The pro rata reduction has a certain amount of precedent behind it, the world community having invoked it in many areas from arms to sulfur dioxide (SO_2).[5]

But it is important to appreciate that as a reduction strategy, pro rata cutbacks are flawed both ethically and economically. Ethically, the objection is that they put a "freeze" on the relative degree of pollution and development, safeguarding the developed countries' industrial advantage while mooring the LDCs to the bottom.[6]

As a matter of economic efficiency, a mandatory across-the-board cutback fares no better. Such a plan, whether expressed in absolute or percentage terms, but in either event indifferent to the diverse marginal costs of the actors, is blatantly and demonstrably inefficient. To illustrate, suppose that the Rhine negotiators are determined to eliminate five tons of chloride emissions per minute. Assume hypothetically that all five nations inject the river with equal quantities. Of these, one country, let us imagine Germany, has already unilaterally clamped down on its polluters to the point where further reductions will cost it $100 a ton. By contrast, another country, say, Switzerland, has failed to implement the most modest—and cheapest—measures.[7] Switzerland is still dumping into the river chlorides that could be eliminated at $20 a ton. Surely, if the community as a whole were determined to eliminate five tons it would reject as inefficient an order that these two countries eliminate one ton each at a cost of $120 (Germany's $100 plus Switzerland's $20) rather than to arrange for Switzerland to eliminate both tons at a cost of $40 (two times $20).

Finally, note that the thrust of this argument goes only to the question of efficiency: the pro rata cutback is inefficient. But if the efficient, $40 solution is selected, there remain further questions of justice. Should Germany, which would have been required to expend $100 had the reduction been mandated either on a per nation or pro rata basis, now be asked to shoulder all or part of Switzerland's expenses? There are in fact several intuitively plausible approaches to the distributional questions.[8]

Whatever the actual facts on the Rhine, the issue is clearly a critical one in the climate change negotiations. Japan in particular has already squeezed much of the easily eliminatable carbon out of its

133

economy, to the point where further reductions on a ton for ton basis are much more expensive for Japan than for other nations. Clearly, if cutbacks are to be made in global CO_2 emissions, it would be more efficient—the global economy would be healthier overall—if cutbacks were emphasized in other, laxer countries. We do not know precisely and directly how much a carbon reduction would cost each country at this point. But we can get a pretty good idea by ranking nations by reference to their CO_2 emissions relative to their gross national product (GNP), that is, the carbon pollution it causes to produce $1 in goods and services. The results are striking (see table 2).

What the table shows is that Japan and France (the latter heavily reliant on nuclear energy) are the most carbon-efficient. China, South Africa, Romania, Poland, and India are the most probable "soft" spots. In other words, if our hypothetical Authority were looking for areas where further reductions in carbon output could be achieved most cheaply, it is to those areas it would turn first. Of course, in the actual world community, these are nations no one is going to come down hard on. But it does suggest this: *the cheapest way to remove a ton of carbon from the atmosphere would be to transfer technology to the countries with dirtier production techniques.*[*]

EFFLUENT TAXES

The more ideal (perhaps idealistic) approach is the "pollution tax." We do not want firms to do $1 million worth of pollution damage if it can be eliminated with a simple $1,000 investment. On the other hand, we do not want the firms to spend $1 million dollars in abatement, spiraling the costs of their goods and services, if their doing so will yield the public only $1,000 benefit in cleanliness. At least at a

[*] And recall from chapter 3 that this echoes the tactic West Germany adopted prior to unification when it paid the lion's share of the costs for cleaning up the run-down Buna chemical complex in East Germany: Bonn had reached a point where it could get more domestic cleanup from a Deutschemark spent abroad than from one spent at home.

TABLE 2

National Carbon Dioxide Emissions Related to National Economics

Country	Emisssions (metric tons CO_2/year)[a]	GNP (billions of $/year)	Emissions per Caput (metric tons[b] per caput	Emissions/GNP Ratio (metric tons CO_2/year
China	2,236.3	372.3[c]	0.5	6.01[d]
South Africa	284.2	79.0	N/A	3.60
Romania	220.7	79.8[c]	N/A	2.77[d]
Poland	459.4	172.4	N/A	2.66
India	600.6	237.9	0.2	2.52
East Germany	327.4	159.5[c]	5.4	2.05[c]
Czechoslovakia	233.6	123.2[c]	N/A	1.90[d]
Mexico	306.9	176.7	1.6	1.74
USSR	3,982.0	2,659.5[c]	3.8	1.50[d]
South Korea	204.6	171.3	1.3	1.19
Canada	437.8	435.9	4.6	1.00
United States	4,804.1	4,880.1	5.3	.98
Australia	241.3	246.0	4.0	.98
United Kingdom	559.2	702.4	2.7	.80
Brazil	202.4	323.6	0.4	.63
West Germany	669.9	1,201.8	3.0	.56
Spain	187.7	340.3	1.3	.55
Italy	359.7	828.9	1.7	.43
Japan	989.3	2,843.7	2.2	.35
France	320.1	949.4	1.6	.34

[a] Source: National Academy of Sciences *Policy Implications of Global Warming* (Washington, D.C., 1991).

[b] Source: T. A. Boden, P. Kanciruk, and M. P. Farrell (1990). *TRENDS '90: A Compendium of Data on Global Change.* Carbon Dioxide Analysis Center, Oak Ridge National Laboratory, Oak Ridge, Tenn.

[c] Estimates of *GNP* for centrally planned economies are subject to large margins of error. These estimates are as much as 100 times larger than those from other sources that correct for availability of goods or use free-market exchange rates.

[d] The emissions/*GNP* is also likely to be underestimated for centrally planned economies.

theoretical level, it has long been recognized that the "right" balance between pollution emission and pollution abatement could be achieved if the firm were charged a tax that confronted it with the full social costs of its pollution. If the true cost of a ton of salt dumped into the Rhine is $100 (measured in reduced amenities, destroyed aquatic life, damages to downriver tulip growers), then each firm should be forced to pay $100. Under the threat of the $100 per ton fee, the firms would be forced to make the socially "right" investment in pollution control, neither too much nor too little.

The tax overcomes many of the objections to the "polluter pays" lawsuits. While collective goods, such as the value of a clean river, are not accounted for in the ordinary court suit, the tax can adjust for them. Because, unlike the ordinary "polluter pays" suit, it is not the victim that collects money from the polluter, but the taxing authority; the victim retains the incentive to take efficient defensive measures.[9] And the tax can be calibrated to look beyond the immediate "damages" that a court would award a plaintiff to account for the full long-term social costs and benefits to everyone.[*]

The fact that the tax does not track down specific victims to compensate, the way civil suits try to do, may be regarded as a drawback. But if the tax is set right, so that the level of pollution is appropriate, there is no "victim" meriting compensation. Moreover, in situations such as along the Rhine, where there are many contiguous polluters and many widespread victims, civil suits are not especially satisfactory, anyway.

Despite all that can be said in favor of the taxes, they, too, have serious drawbacks.

To begin with, there is the problem of setting the "tax" at the right level—just equal to the social damage it will cause in the long term

[*] Suppose that the salt dumping is causing social damage of $100 per ton at present. But in the long run, after both the potash mines and the tulip growers have found the optimal combination of restrictive and defensive measures (the mines undertaking some land storage of chlorides upriver, the growers some desalinization of water downstream), the marginal damage would be only $20, then the right tax is one based on the $20 externality. A tax of $100, corresponding to the current damage, would be overrestrictive.

when polluters and victims are responding optimally. If the tax is too low, there will be too much pollution; but if the tax is too high, there will be too little production, that is, socially beneficial activities will be closed or curtailed. It is therefore critical to determine what the true long-term social cost is of a ton of chlorides dumped into the Rhine, or of SO_2 vented into an air basin. The only honest answer is that it is hard to say.

Take SO_2. When we get down to it, we have far too little idea of its effects, either on human lungs or, through acid rain, on lakes and forests, to put the harm into a dollar value as tax. The best we can venture is that the amount and sort of damage depends on circumstances: where and at what altitude and under what conditions the fumes are released. A ton of sulfur released into a residential air basin presumably does more damage than a ton vented over a sparsely occupied mine site. But that makes a flat-fee emissions tax inefficient, for if we charge the unified rate on everyone's ton, disregarding the circumstances, some polluters will be charged too much (more than the damage they are doing) and others too little.*

Moreover, while taxes appear to be less costly to administer than complex multiparty lawsuits, they are not inexpensive to administer *well*. If the flat tax, by overlooking variations in individual polluters, is unacceptably distorting and we wish more closely to calibrate each emitter's tax to the damage it is doing, then an accounting has to be made of the right level for each point source or area. And even when the approximately right tax is set, we have only shifted from the courts to the tax collector the burden of determining who is discharging what.

In the case of automobiles, the Environmental Protection Agency (EPA) test-drives new models and imposes a graduated excise tax on

---

* Moreover, leveling a flat tax on a pollutant works best if the damage attributable to a pollutant is flat—in other words, if each ton of x discharged into a river causes $10 damage over every range of output. Often, though, the damage is fluctuating, even discontinuous: over some range of discharge the damage is negligible; then, with additional levels of pollution, the harm "spikes"; and then, beyond that level, additional pollution makes little difference because all the fish are dead, etc. This, too, complicates establishing and administering the ideal tax level.

any manufacturer whose fleet fails to achieve overall fuel economy of approximately 27.5 miles per gallon.[10] The technique is rough justice at best, both since the social damage from a pound of emissions is contingent on locale and the actual amount of driving any car owner will do is uncertain. Nonetheless, with mass-produced goods, there is a certain pragmatic virtue in monitoring at the point of production and levying a tax at the point of sale.

When it comes to monitoring effluent from a legion of dissimilar production facilities, however, the tax authority is faced with a harder problem. In general, the more pollution sources there are to audit and the more the auditors are committed to reflecting in the tax the actual damage caused, the less viable the system.* This partly explains the experience of the Netherlands, where a pollution tax on industrial waste-water discharges, in force since 1970, is reportedly a success, whereas market-oriented efforts to restrict agricultural pollutants, for which there are perhaps 100,000 sources, has been far less satisfactory.[11] There are simply many more farmers than factories—more to monitor, and more to fend off in political forums.

The Special Case for the Carbon Tax

The worldwide publicity that proposed carbon taxes are drawing merits some special attention. From at least one standpoint, a tax strategy in the climate context is easier to defend than taxes in the more familiar contexts of domestic water or air basin pollution. As we saw, in those situations taxes were complicated by the fact that damage from the same chemical agent can vary considerably, depending on the point of release—whether upwind or upstream of a large population center, for example. By contrast, the risks of ozone-depleting agents and GHGs are roughly cumulative and independent of source. In fact, as an element in the phasing out of ozone-depleting

---

* Although defenders of the effluent tax can certainly point to the fact that the very reasons that make monitoring and collecting of taxes problematical will probably make an alternative system, such as fine collection, difficult as well.

chemicals (ODCs), the United States has slapped an excise tax based on each ODC's calculated "ozone-depleting factor."[12]

In many ways, unfortunately, setting the right level of carbon tax is more difficult—even more than in the case of ODCs, which are on their way out anyway, under the ozone protocols. In the more familiar contexts in which effluent taxes have been discussed, in regard to smog in air basins or levels of mercury in streams, we know that reductions are of *some* benefit at *all* margins to *everyone*. But when it comes to constraining GHGs, there is even some conceivable net benefit from some degree of warming. This leaves us with an even poorer grasp of what we need to know: the costs and benefits, specifically when all societal actors—generators and victims—are behaving at the optimal equilibrium. But who knows the marginal social cost at current, much less at future levels of activity, when our descendants will be deploying technologies we today can barely imagine?

Although our knowledge of costs and benefits is imperfect, if we are persuaded that some classes of GHGs were on net detrimental, we could impose a carbon (or GHG-indexed) tax believed to be safely within the range of our best estimates. In other words, suppose that while there no agreement could be reached as to whether $20 or $30 per ton was the right tax for gas x, we could say with a high level of confidence that $15 would not be an overtax. Even with that limited aim, could an acceptable scheme be built around a $15 tax, at least as a partial response?

The answer is: perhaps. And there is every reason to press forward and examine the prospects. But the problems facing negotiators cannot be overemphasized.

To begin with, the proposals to tax can be gathered into at least two rough headings. In the first group, covenanting nations would agree that each would levy the appropriate tax on its own domestic carbon (or GHG) use, the funds to go to each government's own domestic treasury.[13] Under the second approach, the tax would be levied under the auspices of some international authority, the proceeds to be available for international uses.[14]

A defect of the first approach is the likelihood that each nation, wary of competitive disadvantage, will be timid in its application and collection policies, without assurances that competitor nations will sedulously carry through on comparable taxes. Opportunities for conflict are rife. Around the world, governments seek to boost industrialization by subsidizing carbon use. It has been estimated that the Indian government subsidizes coal at $8 per ton, China at slightly less.[15] A government determined to press ahead with development could overwhelm an agreed-upon tax by raising the subsidies. As we know from experience with antidumping laws, identifying and challenging such subsidies can be an administrative nightmare.

Second, there is the risk of "addiction." As governments become reliant on the carbon tax as a source of funds, they may become ambivalent about eliminating carbon use—just as local governments that come to rely on cigarette and liquor taxes lose incentives to discourage their consumption.

The second mode, the tax imposed by an international authority, avoids the addiction problem. Because the tax is taken out of the national government's hands and turned over to the United Nations or some comparable agency, each government has the right incentives to bring its polluters to heel. Yet even then the international tax would be imposed upon nations, not firms. There would therefore be no assurance that after the national tax bill had been run through the domestic political machinery, the tax would be driven home as an added cost on carbon users.[16]

Perhaps the most fatal drawback of taxes of either sort is that there is little hope that GHG use would be appreciably dampened by taxes anywhere near the bounds of what the marginal costs, most robustly estimated, would indicate, or what the political system, at its most pliable, would likely accept. Because energy costs are only 3 to 5 percent of the costs faced by modern European industries, a surtax on carbon content of even $10 a barrel (a proposal currently running into stiff opposition in the European Community) would have minimal effects on fuel use.[17] One EPA study concluded that even a tax-driven tripling of carbon prices worldwide would delay a 2° C rise, if that were in the offing, by only about five years, from 2040 to 2045.[18] The U.S. Department of Energy (DOE) recently analyzed the feasibility of carbon taxes as a device for reducing domestic CO_2 emissions to 20 percent below 1990 levels by the year 2000. The DOE concluded that a tax of $500 per metric ton would be needed, more than doubling the price of gas, heating oil, and electricity, and costing consumers $95 billion a year.[19]

The bottom line is that we appear to be caught on the horns of a dilemma. On the one hand, taxes that fall within a politically realistic range would likely make no noticeable difference. (Of course, if the long-term marginal benefits of abatement are insignificant, the taxes should be at a level that does not make much difference!) On the other hand, if it turned out that the right carbon tax was high enough to dampen use appreciably, the revenues would run on the order of hundreds of billions of dollars per annum. As we know from Rio, this is far more than the major nations would be prepared to turn over to the United Nations or to any other world agency for disposal.[20]

TRADABLE EMISSIONS PERMITS

Of economists' proposals to harness market forces, the creation of markets for emissions permits is the most prominent and imaginative. To understand the concept, let us examine it first in a domestic application. Imagine an air basin into which 10,000 tons of volatile organic compounds are being pumped yearly. The EPA declares the region in "noncompliance" with Clean Air legislation. As a consequence, a freeze is placed on levels of emission.

The question then is how to assure that, within the 10,000 ton limit, the region's output is most efficient. Of course, having no air pollution at all would be marvelous. But assuming that some industrial activity is to continue, the best we can hope for is assurance that the pollution that does take place within the constraint gives us the most social benefit per pound of pollutant. In other words, given the choice between two firms, one whose pound of pollutant comes from an activity that yields $10 profit, the other whose pound yields only $5, we would normally prefer—forced to the choice—the former to flourish at the expense of the latter.

141

That is roughly what the marketable emissions approach aims to achieve. Under such a plan, each of the major polluters in the basin would be issued permits to pollute up to a certain share of the aggregate 10,000 tons allowable. How the original allocation of rights is to be made is a separate and controversial matter; rights could be distributed on the basis of past usage, or they could be auctioned (see below).[21] By whatever standard the original distribution is achieved, from that point on each holder of a right can pollute up to the amount specified in its permit.

That does not leave the firms without incentive to cut their pollution. On the contrary, any firm that can devise an improvement and pollute less than its authorized limit gets an Emission Reduction Credit (ERC) which it can either bank for future expansion or sell to others. If a company wants to pollute more than its authorized limit, it has to purchase an ERC from another polluter willing to cut back its own pollution.

For example, suppose New Co. appears, wanting to set up a new plant that will increase employment and wealth in the region, but which will also add 1,000 tons of hydrocarbon. New Co. cannot set up shop unless it arranges an "offset," that is, pays other polluters in the basin to cut back their hydrocarbon emissions by at least the 1,000 tons that it plans to add.* In areas where such trading is instituted, markets in the rights even develop to help clear the required transactions. In Los Angeles, ERCs of various sorts were selling in 1985 for between $850 and $5,500 per ton per year, depending on the pollutant.

Ideally, the regional pollution declines to the targeted level, and does so in a way that apportions the onus of reduction far more efficiently and fairly, and with less administrative cost, than a com-

---

* Offsets may also be employed. Suppose that Existing Co. has a large complex facility that emits pollutants at various points. It wants to make a shift in production that, while reducing pollution at one of its buildings, will increase pollutants at another building in the same air basin. This is presumptively unlawful ("no new sources"). Nonetheless, by conceiving of the whole facility as a single "bubble," a sort of internal trading is allowed: the increase at one portal is permitted if it is linked with a compensating decrease at another. What results is more social value, with no additional pollution.

mand-and-control approach under which a central authority mandates which firms shall abate how much and by what means.

Indeed, both the basic aim and the underlying wisdom of the emissions permits are much the same as those that underlie the pollution tax. With this difference. Under the emissions tax approach, the regulators start by estimating the marginal costs and benefits of some form of pollution and its abatement, and levy a tax accordingly. The tradable permit procedure starts off with the regulators estimating a target *quantity* that they think is within tolerable limits of health, etc., and that they hope can realistically be achieved. Under the tax approach, if it is working well, the regulators will get firms to put into pollution control the *costs* the regulators want invested. They are not apt to wake up one morning and find they have busted the economy. But if they set the tax too low (because, say, they underestimate the health damage or overestimate the costs of containment) there will be too much pollution—much more than they would have tolerated under a permit approach. On the other hand, if the permit approach is adopted and working as it should, the regulators will get the *quantity* of pollution they are willing to countenance. But they may discover to their regret that the costs of achieving that target are far more onerous than they would have mandated under a tax approach.

In theory, the choice of instrument depends largely upon the quality of information available to the regulator and, in particular, to the most likely type of regulatory error.[22] As a practical matter, if the negotiators have a high degree of confidence about the "right" level of emissions—for example, if they believe there is some real ceiling beyond which the risk of bona fide calamity increases exponentially—then it makes sense to address quantity directly, through permits, rather than try to induce the desired quantity indirectly, by guessing at the "right" fee.[23] The tradable permit system provides the least-cost way to achieve that politically agreed-upon target,[24] inasmuch as the emissions that do take place represent the social optimum: those who can make the most of the allowable units buy them from those in whose hands they are less productive.[25] The system is thus markedly superior to the widely promoted mandatory pro rata cutbacks, discussed above.

143

Thoughtful proponents of marketable pollution rights claim the system, applied domestically, has the potential to save billions of dollars annually nationwide.[26] Some regret has been expressed that the most recently proposed southern California plan appears to revert toward heavier reliance on command and control techniques.[27] Yet that backslide is instructive: it appears to reflect concern that at some stage, as the number of point sources and pollutants subject to the regulation increases, a tradable permit system may be simply unpoliceable. How can any administrator be certain that relatively small-scale sellers of ERCs really have earned the credits they are selling? Or verify that they have really cut back on their own pollution after the sale?

On the other hand, the supporters of the idea contend that despite some scattered experience with emissions markets in the United States, none of the variants they would consider "ideal" has yet been attempted. There has been relatively little interfirm trading, in particular. This has been attributed to the fact that most of the efforts to date have accompanied regulations too stringent to leave many credits available for interfirm trade. Another explanation is the difficulty of eliminating uncertainties about the value of the supposed "property right" being offered for sale. A would-be buyer cannot know the worth of an ERC when political administrations change. This affects the "banking" feature also. Companies reducing pollution have never had complete assurances that credits they "banked" from present reductions would be respected five years down the road rather than taken as proof that further eliminations were practicable under a newly imposed, more stringent successor regulation.

Domestically, the most important test is imminent. The Clean Air Act requires electric utility companies to cut their SO_2 emissions in half by the year 2000. Allowances will be distributed in March 1993 in denominations of one ton of sulfur per year. Companies that emit less than their allowance will be allowed to trade the unused portion.[28] Companies that exceed their allowance will be penalized $2,000 per excess ton.[29] It may work. The Chicago Board of Trade has been designated by the government to run the world's first exchange in pollution credits. Even in advance of allocation and the establishment of the Chicago market, utilities were engaging in futures contracts. In May 1992 the Tennessee Valley Authority purchased enough allowances to emit 10,000 tons of SO_2 from Wisconsin Power & Light, a privately owned utility, at an estimated cost of $2.5 to $3 million ($250 to $300 per ton).[30]

Emissions Trading in the Global Context

Even before the jury has returned on the domestic experience with permit trading, commentators are already beginning to tout the approach for application on the global level.* This is not surprising. We have already seen that the atmosphere, ozone shield, and oceans can well be regarded as common resources, like an unfenced commons open to all users without charge, and therefore at risk of being "overgrazed." The idea behind a worldwide system of tradable permits is to establish private rights to certain uses of the commons areas; principal focus is presently on the atmosphere's CO_2 and ODC storage capacity, but tradable permits can also be considered for ocean dumping and deep-sea fishing (see chapter 9). Enforcing tradable rights would be aimed at raising funds (at least if there were a start-up government-conducted auction) and at achieving the efficient level of use of commons areas—or, at the least, to sensitize users of the commons areas, the carbon emitters, ocean dumpers, and others, to the social costs of their exploitation.

Focusing on proposed CO_2 plans, the covenanters would first negotiate an agreeable level of total carbon (or all-GHG) emissions, and then create and allocate as property rights marketable portions of the total. Compliance could be secured by fines, the expected level of which would obviously place an upper limit on

---

* There may also be an important intermediate role for emissions trading: as a way for individual nations to meet their commitments under internationally agreed constraints. In 1988, when the EPA first published its proposed final rule regarding CFC phaseouts, it announced its intention to allocate production and consumption rights based on 1986 levels (called the "allocated quota system"), but at the same time solicited comment on whether to supplement or replace the allocated quota system with an auction system or a regulatory fee system. However, since then the EPA has allocated production and consumption levels without any further mention made of the auction system in any formal notice.

the price any would-be exploiter would pay for the permits. Fines for unpermitted use could also be one source of an international fund which, among other functions, might be empowered to buy up carbon permits and thereby accelerate the reduction schedule if the original pace appeared, with experience, to be inadequately restrictive.

There are at least two thoughtfully outlined proposals for a globalization of carbon-trading rights, one by Michael Grubb and the other by Joshua Epstein and Raj Gupta. Both score many telling points; but they also reveal several drawbacks which, while not fatal, would have to be overcome.

The difficulties of the schemes include, most obviously, designing an institutional structure that can manage the permits, monitor compliance, and provide permit traders and bankers with the necessary assurances that the system will remain stable over time.[31]

But the problems go even deeper. Trading presents the other side of the ignorance that vexed establishing an ideal tax. Just as we do not know the right "price" (effluent tax) with which to confront polluters, we are not even close to agreement on what is the right quantity of emissions. That uncertainty is a large part of what the greenhouse controversy is about. If we allow too much GHG to build up, we face the risks of intensified climate; if we clamp down too hard and too fast, we face the risks of the other side: worldwide economic disruption, unemployment, and civil unrest.

Worse, even if we could come to an agreement on the permissible aggregate, the initial allocation is a hornet's nest. Suppose the permits were allocated on the basis of willingness to pay, via an initial auction conducted by the United Nations. Some would complain that "willingness to pay" means "ability to pay" and that the plan therefore favored the rich. Even aside from that objection, recall the criticism of the global carbon tax, that it would be politically unacceptable to place at the disposal of a world body the amount of revenues that would be raised by a significantly demand-dampening levy. Auctioning methodologies are tricky, making predictions of outcome difficult. But it is not clear why a permit-driven restriction of comparable magnitude, auctioned by an international agency, would not produce the same value as the carbon tax but in one rush:

roughly, a lump sum equal to the discounted value of what the tax stream would yield over the space of years.

There are allocation formulas other than willingness-to-pay. Rights could be apportioned by reference to gross national product, to need, or to historical patterns of usage (thus giving existing polluters a prescriptive right). Even territorial size has been proposed.[32] Each has obvious theoretical and diplomatic drawbacks and few supporters (although, at least in domestic application, allocation according to the historical pattern has a cagy political advantage: giving priorities to existing polluters is a nice and perhaps necessary quid pro quo to secure the support of those with the most power to block restrictive legislation).

Both the Grubb and Epstein-Gupta plans select national population as the baseline. In the Epstein-Gupta version, the permit regime would first be divided into accounting periods in which each nation's population would be projected. Second, a global emission target would be established for each period. Next, each nation would be assigned a pollution entitlement as a simple proportion of its share of population to the global target. Finally, each country's "laissez-faire emissions" would be identified—its projected emissions if there were no constraint. If a nation's allotment exceeded that level, it could sell permits for the excess to others; if otherwise, it must reduce or purchase.[33]

An advantage of the proposal is that by providing for direct nation-to-nation trading—for which there is even some precedent in the ozone convention[34]—controversy surrounding overfunding of international agencies can be skirted. Moreover, one awkward drawback of a permit system indexed to population—that it undermines population control incentives—can be finessed by indexing to adult population or population at the time the treaty enters force (so that a nation acquires no additional entitlement by adding births).

But there are other problems not so easy to dispel. One is calculating the laissez-faire level: How are negotiators expected to get agreement on the amount each nation would emit if it were unconstrained? Moreover, if the trading scheme is structured to provide direct nation-to-nation allotment sales, it could entail a transfer of wealth too colossal to be acceptable politically and too uncertain in its consequences to be supportable theoretically, politics aside. The less populous, highly developed countries—the United States, Japan, the European Community—would not all of a sudden halt their carbon use while it just as suddenly burgeoned in India and Africa. Depending upon the aggregate emissions allowable (which could be as ambitious or as cautious as any tax plan), patterns of fuel use would continue as much as possible as before, with the principal difference being that the industrialized world would have to transfer wealth to the LDCs under the guise of buying the latter's unexploited share of the "privilege" to pollute. The newly enriched LDCs would then be positioned to industrialize—and, unless historical patterns are dramatically snapped, increase their carbon use. When they ultimately reached their own limits, could they be counted on to buy back the rights they would then need from the industrialized countries? Or would they find an excuse to withdraw from the system?

The idea of improving the environment through internationally permitted GHG trading is a good one and deserves continued exploration. Delightfully promising illustrations crop up. Consider this. Each wild salmon in eastern U.S. rivers is valued at $500 to $1,000 because of its contributions to the sports fishing industry, where the fish represent profits for tourism, motels, tackle shops, car rentals, and so on. Many fish that might reach the U.S. rivers to spawn never get there because they are intercepted on their migration by fishermen off Greenland and the Faroe Islands. Yet each fish is worth only $15 to the fishermen. The proposed solution: the North Atlantic Salmon Conservation Organization (NASCO) allots salmon quotas to participating members. Private groups have initiated plans to raise a fund to buy out the Greenland and Faroe Island quotas, compensating the fishermen for the profits they will forgo as the fish pass safely westward to the United States, and also underwriting alternative job training for the fishermen. It is suggested that, if NASCO agrees to make the quotas tradable, a $1 million fund to purchase quotas would go further than the $35 million the United States is currently expending each year on domestic steps to restore wild salmon, principally through river repair. Similar plans aimed at easterly movements of the fish have doubled the number of fish returning to European waters.[35]

Other twists are conceivable. There is opposition to catching tuna "set" on dolphins because dolphins get entangled in the nets and die. The new tuna convention agreement Inter-American Tropical Tuna Convention (IATTC, June 1992) provides that the current "kill" of dolphins cannot exceed 19,500 and has to decline to 5,000 by the year 1999. The technique is for IATTC essentially to allot to each fishing nation a per vessel share of the constricting Eastern Pacific quota: an allowable number of dolphins that can be killed per boat. During negotiations, there was some discussion of making the kill quotas marketable, so that pro-dolphin people (and friends even of tuna) could buy up a boat's right to kill dolphins. The proposal was dropped in face of opposition by strong environmentalists, led by Greenpeace, who maintain that a dolphin has a right to life that cannot be bought and sold—and that nothing short of zero kills will be acceptable.

There are undoubtedly many opportunities of various sorts in which tradable permits of more modest ambitions than global carbon trading can have a positive impact. But it is important to remember that, by itself, trading will not eliminate or even curtail pollution and other degradations of the environment. What tradable permits may offer is a relatively efficient means of achieving a target that has been independently set. That is not an insignificant prospect; indeed, the more efficiently and fairly pollution can be eliminated, the more feasible it becomes politically to establish ambitious reduction targets. But all the truly global schemes advanced thus far raise serious questions for which neither domestic experience to date nor sober horse sense provides reassuring answers.

Conclusion

Despite the difficulties of identifying efficient outcomes, the principle is worth remembering: in a world in which so many desires are unfulfilled, and in which there is so much poverty, we cannot regard the environment and its amenities to be infinitely valuable; nor can environmentalists muster political support if they regard each and every response to each and every crisis as equally meritorious.

Hence, the costs and benefits of environmental protection—as best we can estimate them—have to be compared constantly with competing demands. Economic analysis provides the best single way to keep these compromises in view. Indeed, it is a shame that economic analysis is so commonly disparaged by environmentalists, who have somehow gotten the idea that economic thinking and environmental thinking are inherently opposed.

Environmentalists often feel that, by admitting there might be a marketable "right" to pollute, they have given away the store. But some compromises are necessary, and making polluters pay for polluting—thereby reducing the levels—is a whole lot better than allowing them to get away with it for nothing.

The more durable objection is that many environmental values, such as the "value" of noncommercial species, are not captured by markets. As Mark Sagoff has argued so spiritedly, the value placed on preservation has to be set in the political arena, not in markets. "The genius of cost-benefit analysis," Sagoff remarks, is to inhibit "conflict among affected individuals from breaking out into the public realm"—which is exactly where many of them ought to be.[36] But this "market" objection goes only to the most cramped conception of economics. Economic analysis is certainly robust enough to accommodate nonmarket-measured values, although it takes an alliance with the legislature: as a first step, the public bodies can ask citizens what they are willing to pay to preserve a noncommercial species or exotic habitat, and base social decisions accordingly.*

Let me illustrate with the question whether to permit or even encourage wolves to reestablish in national parks. Wildlife groups favor their return. The stockbreeders on the fringes of the parks are against

---

* There are three methods for conducting such surveys. One is to seek out an *existence value* by asking: "What is it worth to you just knowing (having the assurance) that bowhead whales exist?" The second is to determine *option value*: "You may not want to travel to Alaska to watch the migration of bowhead whales now, but what is it worth to you to keep that option open?" The third aims for *bequest value*: "Just as it is worth something to you to defer spending some of your fortune so as to leave it to your children, how much is it worth to you to leave bowheads to future generations?" Unfortunately, all three methods are susceptible to a wide range of objections; many commentators believe there is a tendency for those questioned to

it. The ranchers argue that the wolves eat their calves. The conservationists say the ranchers exaggerate; and, anyway, why prefer calves to wolves (or why prefer people eating calves to wolves doing so)? Around Yellowstone there was a virtual standoff—until Defenders of Wildlife came forward with the idea of funding a trust to indemnify the cattlemen for any cattle they lose. In 1992, as a more positive incentive, an additional offer of $5,000 was made to any private landowner whose land hosts a den producing pups allowed to grow to maturity.[37] The compromise solution was an idea that any economist would be pleased to take credit for, and one that shows the potential of "economic instruments" more modestly scaled than global trading.

But closer analysis showed even more. The losses to the state of Wyoming in hunting revenues (stemming from the wolves' predation on moose, elk, and deer) was on the order of $500,000. When people, in particular prospective park visitors, were asked what they would be willing to pay into a hypothetical trust fund to see wolves in the wilds, it turned out that the decision to reintroduce the wolves may have been worth $30,000,000[38]—again, the "right choice" on simple economic grounds, even before we take up the deeper question of an environmental ethic.*

Here, the point is simply this. The mutual distrust between economists and environmentalists is unfortunate. I agree with Amartya K.

---

overstate their true valuation of environmental amenities. See Note, "'Ask A Silly Question . . .': Contingent Valuation of Natural Resource Damage," *Harvard Law Review* 105 (1992): 1981–2000. Other commentators suggest that many respondents, when asked how much they would pay for a threatened environmental amenity, *understate* their true valuation because, misunderstanding the thrust of the question, they believe some villain—an oil company, not they—ought to do the paying. But nothing is sillier than to cite these flaws in the surveying techniques as an excuse for ignoring the question of nonuse values entirely. Weak data critically analyzed are better than no data at all.

* In other words, suppose that the amount of money that conservationists and park visitors are able to come up with falls short of the prospective economic losses to the ranchers and hunters. Does that end the argument on behalf of the wolves' return, or is there something more to be found in a duty to Nature? See chapter 10.

Sen that bridging the gap between economics and ethics would operate to the benefit of each.[39] Economic thinking and economic instruments, with whatever occasional shortcomings in their assumptions about human motivation, keep us continuously vigilant as nothing else does to costs, benefits, alternatives, and opportunities for gains from trade. However, they will not provide us, as much as we would like it, the grand solutions that go right to the heart of our problems. We don't have the data that economics needs, and economics, even broadly understood, cannot arbitrate all of the conflicts we as a community want resolved. In casting about for solutions we should not overlook the contributions that effluent taxes, tradable allowances, and the homier class of tactics exemplified by the wolf fund. But for all their elegance and promise of apparent simplicity, we will have to look beyond the economic tools to chip away at the problems with a full chest of readier, rougher, more varied, time-proven and modest devices.

CHAPTER VII

Medicating the Earth: Preventatives
and Remedies

T HE FACT that economics cannot provide a nostrum for environ-mental ailments should hardly be surprising. The planet's maladies are too complex and its cultures too diverse to rely on any single line of attack. Ideally, we should be aiming for a taxonomy of global-degradation problems that can be matched with a taxonomy of con-trol strategies. That is, what sorts of institutional responses appear best suited to what sorts of environmental problems? One imagines that some challenges would lend themselves to relatively strong reli-ance on economic instruments, such as excise taxes; for others, such as ultrahazardous chemicals, we best shift to preventive standards such as mandatory safety devices; for yet others, including some of the less well defined international conflicts, the most we can hope for may be some institutionalized patterns of notification and consulta-tion among diplomats.

A thoroughgoing, problem-by-problem taxonomy would be, as I say, "ideal." But it is also slippery. Typically, the characteristics, both of the problems and of the control responses, are diverse and still ill-understood. How do we know what instruments to call upon to combat red tides, if we do not know what is causing them? How can we deploy a carbon tax, when we are so uncertain about both the social costs of alternative fuels and the environmental damage car-bon causes? Some remedies are highly situational. Attaching con-ditions to debt relief may serve as a lever against poor borrower nations, but it cannot influence the industrialized lenders.

In fact, the suitability of a remedy may even depend upon how a problem is defined subjectively. If we think of a problem as too-much-dioxin-in-a-river, then there is something to be said for estab-lishing a tight and shrinking market in tradable dioxin emissions permits. But suppose that the local problem is characterized as protection-of-endangered-salmon. On that view, we are forced to consider the interplay of many factors that impinge on the habitat: not merely the dioxin sources (from the riverside wood manufacturing industry, say), but those who use the water, thereby affecting its mass and flow (hydroelectric power companies), other polluters (municipal sewage departments), those who extract water and do not return it to the flow (farmers), and those who crowd out natural tributaries and marshlands (developers). How do we establish a trading plan, or even distribute a tax, among all of them? Indeed, when we regard a problem at that level—as a species in peril—there are not only widely different actors to account for, but already many different agencies in place, each with its own inertia, influence, and style.

That does not mean there is nothing useful to venture about institutional strategies. But many observations have to be confined to a fairly general level. Two popular preconceptions, already anticipated in the preface to this book, serve as a good foundation for an institutional overview. The first preconception is that we are dealing with global problems, and that therefore we are confined to select from a menu of global solutions. The second is that we are always better advised to prevent a problem than to try to cure its symptoms.

Need Our Responses Be Fully Global?

The environmental plight has drawn support to globalism: the credo that the more nations we can get to cooperate in the more comprehensive a global regime, the better. It is easy to see how the idea attracts. Many of the afflictions we have been discussing transcend territorial boundaries and affect people across the world. It is natural to suppose that they therefore require solutions on a global scale: great globe-inclusive multinational treaties, perhaps even a Global Environmental Agency modeled after the Environmental Protection Agency but with earth-spanning powers of investigation and command.

Often this globalist perspective is supported by considerations of legitimacy, particularly where activities on the global commons are involved. Some of the activities we have mentioned, such as seeding space with copper needles, girdling the world's oceans with a sonar blast, or stimulating oceanic algae growth ought not to be done without securing some broad international consensus. Even if such actions are done in the name of satisfying the whole of humankind's safety or curiosity, no one nation should have a unilateral right to put the commons at risk.

And of course the globalist impulse is further reinforced by the evident "prisoner's dilemma" character of many of the environmental problems. In a wide range of cases ranging from ocean pollution to GHG emissions to trade in endangered species, each nation is tempted to exploit its power to the fullest, from fear that if it alone adopts a more cooperative strategy (conserves its elephants, relieves pressure on the fishing stock), other nations will seize upon the opportunity to improve their own welfare at the cooperator's expense. There are thus undeniable advantages to forming an international regime that is broadly inclusive.

During early discussions about CFC reductions, the U.S. government, prompted by American CFC manufacturers, resisted public pressure to cut CFC production through unilateral action. As the leading industry trade journal complained, "If the U.S. takes unilateral action, it takes the pressure off the rest of the world to act" and of course merely disadvantages U.S. industry to the advantage of its European rivals.[1] The ozone treaty—the agreement by *all* major producing nations to reduce proportionately—was a necessary condition of *any* major nation's actions. As in the case of the ozone agents, if enforceable, broadly multilateral agreements can be reached, they are clearly to be preferred.

But we ought not to generalize from these observations to the judgment that global responses are in all instances vital. Try to imagine for a moment the circumstances that would require a global solution in the fullest sense, that is, a peril in which any reaction that did not enlist the cooperation of all the earth's nations would be doomed to failure. Such a case would have to be structured something like

155

this. An asteroid is plunging toward earth. If it strikes any portion of the earth anywhere, it will destroy the planet. Fortunately, technology can be prepared in time to deflect it safely on its way through space. The United Nations agrees to the placement of deflector stations on the high seas and Antarctica, but the technology requires deflector stations to be so spaced that each nation would have to permit the siting of at least one station on its territory. If any nation does not go along with the effort, that would leave a fatal "weakest link"*—a path along which the asteroid will glide to earth and our common doom. In such circumstances the worldwide cooperation of *each* nation would be imperative for every nation's survival.

Or, suppose that one credits the fears that after a certain level of "safe" global warming there may follow a sudden drastic event, such as the collapse of the polar ice caps with devastating implications for coastal states. The catastrophe is not inevitable. It will occur only if all the major GHG emitters persist with their pollution unabated. Put otherwise, an appreciable cutback by any major polluter will, by itself, avoid the cataclysm. Here, in contrast with the asteroid case, a major reaction by any state will save all the coastal states, but not otherwise. In terms of the game-theoretic literature, we would have Chicken, in which each coastal state player would be tempted to continue polluting with the hope that some other "saner" (or simply more nervous) player would "chicken out" and save them all first at the rescuer's expense. But we know that while one nervous country *might* chicken out, there is the awful possibility that none of them will blink in time. In those circumstances, too, the benefits of a truly global agreement merit many negotiating headaches and concessions.

* The story that accompanies the game of Weakest Link (which as far as I can tell is indistinguishable from Stag Hunt) is essentially this: twenty prisoners of war, each of whom knows a secret vital to the enemy's breakthrough of their lines, are being interrogated, one by one, to force them to divulge. If any of them divulges, the valor of those who resisted will come to naught. The group is no stronger than its weakest link, therefore success entirely stands or falls upon universal cooperation. See Glenn W. Harrison and Jack Hirshleifer, "An Experimental Evaluation of Weakest Link/ Best Shot Models of Public Goods," *Journal of Political Economy* 97 (1989): 201–25.

But both the asteroid and the ice-melt illustrations are quite con-trived. In the typical and more realistic circumstances there is little to be gained from constructing an extensive global coalition—and there is a certain amount to be lost. Efforts to cope with land-based sources (LBSs) of ocean pollution provide a good example. The Law of the Sea (LOS) negotiators proposed dealing with LBSs within the framework of a multinational convention open for signature to all the nations of the earth.[2] There are many reasons why the plan stalled. But one explanation, certainly, is that the UNEP Regional Seas Program (see chapter 5), which generates a series of conventions, each tailored to the special needs of countries surrounding a particular sea, met the demand more effectively than a unified "global" solution. Each group of sea-sharing nations typically has a considerable range of shared interests and a common outlook. Why complicate negotiations by inviting participation by other countries, including landlocked na-tions, from across the world?

The United States and Canada have been attacking common envi-ronmental problems bilaterally since 1909, with considerable suc-cess, through their International Joint Commission (IJC) and Great Lakes Water Agreement (1978). The nations abutting the Rhine have been confronting common problems of waterway use since 1804.[3] None of these agreements has been perfect (whatever that would mean). That is not the point; the point is that there is no reason to conclude that they would work better if we expanded them in mem-bership and scope.

Indeed, a certain amount of the support for fully global solutions in this area is driven by people predisposed toward world govern-ment generally, who look to the afflictions of the global environment as an opportunity to further their cause. The same phenomenon was evident in the LOS negotiations, particularly in regard to the then vaunted "wealth" of seabed minerals. The proposed mining regime that emerged was so complex and costly that its principal justifica-tion appeared to be as a laboratory for world government.[4]

The truth is that while broadly global solutions may be useful, and in some circumstances virtually required, they are generally less in-dispensable than one might suppose. The more global any institution, the less flexible it is apt to be and the more difficult to move into action as a body. Agendas clash. Various nations use their votes to extract concessions in their own areas of primary interest. A world-wide meeting of the International Telecommunications Union (ITU), assembled to iron out broadcasting standards, but convened, coinci-dentally, shortly after Israel's incursion into Lebanon, came close to vaporizing over an Arab-state proposal to expel Israel from the ITU.[5] In some circumstances, the mere efforts at global overhaul risk tear-ing the fabric of the budding, more modestly scaled institutions that may handle problems more successfully.[6]

In other words, in the typical situation, broad-scale cooperation offers benefits; but the real—and inevitably open—question is whether the benefits of the broad-scale cooperation are worth the price, including the price demanded by the nations otherwise threat-ening to withhold consent. If the price seems too high, as appeared to be the case at Rio in regard to some proposals, then the alternative of a more modestly scaled regime has to be considered.[7]

I am hardly suggesting that each nation should approach all these problems from the perspective of the most narrowly self-interested calculations of national costs and benefits. As indicated above, many cases, especially those affecting the health and wealth of the com-mons areas, raise legitimate expectations that world community con-sensus will be sought. But not every nation recognizes the need for global consensus even in mediating commons-on-commons con-flicts. For fishing countries, freedom and equality mean that every nation has a right to undisturbed access to take from the commons what it will—you equally with us. Indeed, many of the same tensions we witness in domestic U.S. politics between states rights and feder-alism have close but magnified counterparts in frictions between na-tionalism and globalism—with the further complication that the in-ternational analogues play out in a more loosely knit community that is governed by a unanimity rule. Even in the international arena, too, there is, however, room for innovative institutional compromises. For example, there is a device known as the *dedoublement fonctionnel*, by which international obligations, such as limits on ocean dumping, are delegated to national institutions for enforcement.

Global solutions are often to be preferred; in some circumstances, they may practically be vital. But they have their costs, and, as we shall see, may in some circumstances be less satisfactory than more modestly scaled, even unilateral, efforts.

IS PREVENTION BETTER THAN CURE?

The second preconception is equally critical for the environmental agenda. Environmentalists are increasingly touting "the precautionary principle," meaning: when in doubt, spend now. It sounds right. The trillions of dollars nations have spent on their militaries is proof of the proposition's strength, if not its final good sense. We are, after all, instructed from childhood in dozens of variants that it is better to be safe than sorry, and so on. But if one examines any of these fables and maxims closely (which is not their intent), they say less than first meets the eye. One would indeed rather be safe than sorry, but those who in the 1950s expended many thousands of dollars for backyard "A-bomb shelters" are, I imagine, sorry. And indeed, as foreshadowed in the Preface, that is the problem with preventive regulations—that they carry the risk of overregulation, of making us more sorry than more safe. In fact, risk avoidance has its own risks. By imposing severe fuel restraints that would take a few hundred billion dollars out of the economy we could reduce the risk of greenhousing but increase the risk of massive unemployment and civil unrest.

Anglo-American law, if it is any guide, originated with a wait-and-see rather than a preventive attitude. In general, people cannot go to court to prove apprehension of a risk; they have to wait and show injury. The defect is that in any given case, by the time someone has become the victim of a hotel fire or food poisoning, the ex post remedy appears to have come "too late." But the major alternative, ex ante regulations such as prescribed safety requirements, have the opposite defect: they are deployed before we know which ship on which route is fated to crash from what causes; because each requirement is mandated to apply across the board to an entire class of ships and sea lanes, costs are raised even on the many voyages that would have been plied without incident anyway. How much to lean toward prevention and how much toward cure is a question that, as a general matter, we cannot answer definitively. As we shall see more fully in a moment, domestic law has proceeded to evolve a balance of ex post and ex ante measures.*

THE DEFERRED, GO-IT-ALONE POLICY AS
A SECOND-BEST STRATEGY

When we combine these two thoughts—that neither globalism nor prevention is inevitably as attractive as is popularly assumed—a third insight emerges. Bilateral, even go-it-alone programs that rest in considerable measure on adaptation, are far more appealing than is ordinarily imagined, even for the most facially "global" of problems. This is unanticipated, for while one might suppose that go-it-alone, internal approaches would be appropriate for repair of a nation's interior problems such as urban smog, they would be unsuited for perils that transcend boundaries. But there is nothing inherent even in such a globe-spanning phenomenon as climate change to make each nation prefer to act in concert with others, rather than to mitigate and adapt on its own.

The starting point is recognition that from any nation's point of view, change in the global climate, per se, is never the problem. Aside from the losses of international public goods, which cannot be disregarded, each nation's most immediate concern is some specific and inevitably local manifestation of climate change. As we have seen, these effects may take the form of rising sea levels in one region, loss of soil moisture in another, discomfort in another, and even frost in others.

---

* The prevention versus adaptation discussions are further complicated because some boundaries are blurred. Is planting trees to withdraw CO_2 from the atmosphere preventive or adaptive? Where do we fit cutting off the horns of black rhinos so that poachers will not kill them? Are we protecting rhinos, or—as opponents of dehorning imply, in light of the important role the horn plays in mating, defense, and browsing—making rhinos into something else? See Jane Perlez, "Rhino near Last Stand, Animal Experts Warn," *New York Times*, July 7, 1992, A5.

The implication is that, viewed from many nations' perspective, large-scale cooperative expenditures to prevent climate change from occurring across the globe may be inferior to a policy of adaptation—of waiting to respond in a manner appropriately tailored to its own local problems as they display themselves. By deferring, one does not expend for protection against perils that, with the passage of time and further research, will turn out to have been illusory. And even if perils do eventuate, deferring allows for tailoring. Low-lying nations that find themselves facing storm surges can build storm walls and levees, and so forth. Agricultural regions struck by moisture loss can undertake water impounding and agricultural adjustments when and as required. Moreover, each nation finds it expedient to finance pollution control activities to the point where the marginal costs of its expenditures equal *its* marginal benefits; some nations may choose to stress preventive measures to cut the risks of biodiversity loss and climate change, if that is what suits them. But other, poorer nations may prefer expenditures on child care now, to adapt, if need be, later.

As Thomas Schelling has pointed out, unlike the benefits of typical preventive measures (such as constrained carbon use), the benefits of the adaptive internal strategy are fully internalized.[8] They are therefore easier to motivate. To illustrate, the United States, fearing loss of soil moisture in, say, fifty years, has the option of deferring action (realizing the investment value of the uncommitted funds) and responding with microirrigation, damming, and other water control projects if, when, and where soil dryness becomes a clear threat.* We face a trade-off between carbon constraints now and water con-

---

* When our concern is investments in things such as dams and canals that are characterized by long lives and long construction periods, the options can be intriguing because we have the alternatives to adapt by modifying or even by abandoning the project during construction. In one economic analysis of adaptation to climate change by building a seawall, the most economic course turned out to be to lay the foundation for the wall and then wait to learn more before spending more. Gary Yohe, "Uncertainty, Climate Change and the Economic Value of Information: An Economic Methodology for Evaluating the Timing and Relative Efficacy of Alternative Response to Climate Change with Application to Protecting Developed Property from Greenhouse Induced Sea Level Rise," *Policy Sciences* 24 (1991): 245–69.

straints tomorrow. There is no a priori reason to conclude that a forced dose of the former (a likely component of an international regime) is better for the United States than the latter.[9]

In other words, even suppose that in the best of all worlds, an internationally cooperative carbon-restricting regime *would* be the ideal tactic; in our less than perfect world, in which treaty formation and enforcement are complicated and an effective level of cooperation can be achieved only by unwieldy threats and expensive concessions, a heavy mix of individual, go-it-alone adaptive responses may constitute an important component of a viable "second best" strategy,[10] even for a problem like climate change that seems so facially "global."

A REVIEW OF THE HOME-GROWN REMEDIES

Thus we should press ahead relieved from any preconceived notion that the global environment requires responses that are either uniquely globe-spanning or exclusively preventative. A good way to get a truer fix on the options is through a quick recapitulation of the domestic experience. The correction of "spillover" effects has become recognized as a fundamental task of law within modern American and other societies. A review of those efforts not only provides valuable cues for future domestic policy; it also suggests remedies to cull for possible deployment internationally. Roughly, diplomats working at the global or regional level are substituting sovereign nations for the familiar factories and property owners of domestic law and switching some aspect of the global commons such as the ocean or the Rhine for a domestic public waterway such as the Mississippi River.

There is therefore good reason, both in trying to "export" legal institutions and in designing global strategies, to benefit from the longer history of experience we have had with domestic efforts designed to handle many of the same problems, played out on a more modest stage. I say "roughly" because, as we have seen, international law as it now stands puts barriers in the way of any wholesale transfer of law from the local to the global stage. The global actors, sovereign states, are not just chemical companies writ large. The most powerful oil company can be made to respond to a subpoena and appear in court; not so the weakest nation. Yet, just keeping a range of instruments in mind provides cues as to what legal arrangements we might aim for, even if carrying through would require some reform of international law.

Indeed, as a start, the unfolding of domestic law offers striking microcosmic parallels to what has been going on, with only a later start, on the international plane. At its earliest stages, in preindustrial England, most spillover conflicts were of the simple sort an agricultural society experiences—for example, wandering sheep. Because technology was generally benign and populations relatively sparse, one presumes that modern-style spillover conflicts were fairly infrequent. Thus there seems to have been little occasion to challenge the prevailing sentiment that the rights of property were sacrosanct: each great landowner, in his castle or manor, enjoyed, essentially, a chip of the king's sovereignty. For what you did on your own property, you did not have to answer to anyone.

One must say this was the rule "essentially," because there appeared, even at a fairly early stage, a doctrinal foundation for some modest public review of the property owner's actions: *"Sic utere tuo ut alienum non laedas"*—roughly, "Use your land in such a way as not to injure others." It was a doctrine the early English lawyers could trace back to Roman law, for it must have been clear quite early that one owner's freedom to use his property could have the effect of infringing his neighbor's freedom to use his or her land if the use was abusive.

From this originally frail, sparingly invoked basis for the outside world's complaints, "sic utere tuo . . . ," there gradually developed, over the past century in particular, a whole panoply of domestic constraints on conflicting uses of property.

The earliest actions to evolve were trespass and nuisance, the former for actual physical invasion of another's land (such as causing a rockslide to tumble down on a neighbor), the latter for subtler infringements of enjoyment of one's property (through odors or noise,

163

for example). As we saw, both remedies required the polluter to account for the damages it was (in the commonsensical view) causing the victims and, through them, the society as a whole, and were therefore advances toward the economist's ideal of forcing some internalization of externalities (see Chapter 6).

Yet neither remedy could keep pace with the conflicts of modern life. Trespass was good enough for the small class of truly blatant invasions of land, but not much help in dealing with the growing class of new harms—often sly, remote, and sinister—as industry and populace grew and packed more closely together. To iron out the inevitable conflicts, the courts turned nuisance into something of a catchall. But it too had an instructive defect. Faced with two landowners, both making (or trying to make) profitable use of their land, the courts were forced to mete out justice by trying to decide which party's use was "most reasonable," and what would be the "balance of hardships" if they ordered an injunction to issue.

This reasoning may sound reasonable enough to a layperson (what could be more reasonable than to do the reasonable thing?). But with the passage of time lawmakers began to recognize two flaws at least. First, courts did not always appear well equipped to decide what uses were more reasonable than others—a domestic residence or a halfway house—particularly when both were satisfying societal demands (else why did they come into existence?). Worse, no potential disputants could be certain in advance whether their use was reasonable; it took a court to decide, and the law did not operate swiftly or on the cheap.

The result was that while nuisance standards and other rules of reason are still to be found in various areas of the law, the range of their application has been constrained where possible by more specific rules, often introduced by statute. In fact, there is an old saw about the law—that it is better that it be clear than it be just. By this is meant that if the law at least spells out the ground rules clearly, market values and private transactions will adjust patterns of ownership and behavior accordingly, with relatively little court cost, and even produce, when all is said and done, a system that people will accept as "just." But when the rules are spelled out in terms that are highly indeterminate, many of the potential advantages of a legal system are simply forfeited—indeed, the situation may be worsened. No one knows the rules in advance, and a lot of time is spent in court.

The point would appear too elemental to dwell upon, except that much of international law still suffers from a high degree of indeterminacy, which is unfortunate. The prospect of wildly indeterminate outcomes is one more factor that makes courts reluctant to submit to international adjudication. And it is not good for the courts: if a domestic court is not anxious to step in and arbitrate what is "reasonable" or the equivalent—presumably by reference to a legal and cultural background shared by all the parties—one can readily imagine how much more reluctant will be an international tribunal, already hampered with wariness about its jurisdictional foundation.[11]

Consider, just for an example, the problem of apportioning water among all the many nations that share an international drainage basin. As indicated in chapter 1, conflicts over water usage are potentially one of the globe's most volatile geopolitical issues. But international law has essentially little more to say than that each basin state "is entitled . . . to a reasonable and equitable share."[12]

Of course, if there were enough water to go around there would not be a question of "share." One nation wants to raise barley, the other wants to drive electric turbines, and there isn't enough flow to satisfy both. So, the question is which nation's interests to protect. "Reasonable" and "equitable" are not of much help. Neither is the proposed draft rule that "watercourse states shall utilize [the water] in such a way as not to cause appreciable harm to other watercourse States."[13] Once all the flow is allocated (that is, once we have a genuine problem), then if we let the upstream nation, Upstream, use more water for its barley, it will undermine downstream nation Downstream's plans for electricity. And conversely, if Downstream gets more water and electricity, Upstream gets less barley. This follows from the idea already reviewed: inconsistent users are harming (or are prepared to harm) one another; the harm is reciprocal.

The drafters of such equivocal and toothless rules have to be judged with some sympathy. Water conflicts are intimidating to begin with, and there are no easy answers. Ad hoc multinational drafting sessions are hardly ideal forums for tackling even simpler problems. The United States has not gotten its own water law straightened out well enough to sniff at anyone else's efforts; and, anyway, stern and clear language may be viewed as a potential impediment to collecting the necessary signatures to turn a draft proposal into accepted law. It is natural for treaties to start out as enunciations of vague principle, with the hope that they will mature into detail. (That was the model of the successful series of negotiations on ozone.)

But this judgment may be a mite too generous. One respects a certain amount of flimsy draftsmanship and diplomacy. It is hard to see how those who drew up the ostensible "factors" laid down by the Helsinki Rules thought they were providing any guidance at all. Article 5 of the Helsinki Rules (and in almost identical language Article 6 of the International Law Commission's proposed 1992 draft) provides that what is "a reasonable and equitable share . . . is to be determined in the light of all the relevant factors in each particular case," which are said to "include but are not limited to (a) the geography of the basin . . . , (b) the hydrology of the basin . . . , (c) the climate affecting the basin . . . ,"[13] and so on.

This is not just a case of criterial vagueness; it's worse. To illustrate, when a sentencing judge is told that in meting out a sentence she is supposed to account for the cruelty of the offense or the number of past offenses by the defendant, we may not know how heavily cruelty and past offenses are to weigh in the final analysis (that is the vagueness). But at least we know the direction in which the criteria *cut*: the crueler the offense, the worse the defendant's record, the longer the sentence. By contrast, when one is told to think about the climate, what is one being asked to think *about*? Is the reasonable and equitable solution to give the naturally drier state more water or less?

What are really needed in the area of international nonnavigable water uses are rules and institutions that are less open-ended than nuisance principles and give good clear guidance in advance of litigation or even negotiation. Perhaps here is an area where efforts should be focused on arranging marketable permit systems. Unlike GHGs, each international water basin presents a distinctly regional rather than global challenge, and would therefore seem that much more manageable by markets. True, such a system won't come easy.[14] But this is an area in which the "transaction costs" of the present hazy system may have to be paid on battlefields.

Land Use Planning

In some areas, domestic law's response was to downplay the role of lawsuits altogether and emphasize careful land planning instead. Rather than wait for a conflict to arise and then force a court to sort out the "good" or "reasonable" users from the others, land planning puts everyone on advance notice as to what they can expect. Some of the earliest forms of land planning involved private reciprocal covenants among neighbors. All the homeowners in an area might agree that they, and those purchasing from them, would refrain from putting their land to specified uses such as filling stations.

Today, however, domestic land planning is more often a public responsibility than something patched together by neighbors. Typical devices are regional and municipal planning and zoning boards, whose aim is often not to eliminate undesirable land uses but to restrict it to areas where it will do the least damage and defeat the fewest expectations. Slaughterhouses, for example, one of the earliest targets of zoning, were consigned to one area of the city or banished to the perimeter.

It seems to me there is something to be gained by consciously considering how we might adapt these concepts to the multinational, even global, scale. Internationally, we already have an analogue of mutual covenants among neighbors in the several regional agreements (such as the Regional Seas programs among the nations bordering on the Caribbean and Mediterranean seas, discussed in chapter 4). It is interesting to think about "zoning," too. It reminds us that there is an alternative to banning disfavored activities absolutely; some activities can be restricted in space and time. Biosphere reserves such as those under the United Nations Educational and Scientific Organization (UNESCO) biosphere program can be regarded as existing illustrations,[15] as can the various Antarctic treaties, which spell out permissible and impermissible uses and operations in considerable detail. Ocean-dumping conventions, notably the Oslo Convention, regulate dumping in some areas but others are left uncontrolled. The Indian Ocean has been "zoned" for "no-whaling," and the French are proposing the same status for vast reaches of the southern waters, up to 60° N. Latitude.

But the idea might be extended beyond present usage. Suppose that credible computer models show that wind patterns, etc., made certain locations relatively harmless for certain activities, and other areas less so. Or, it may well develop that, from the whole world's or from some region's point of view, there are some least perilous disposal areas for radioactive and toxic wastes. Might an international (even if regional) convention not do well to think of a covenant or zoning model?

Harm-Based Liability Rules

The earliest antipollution control technique surveyed civil suits for trespass, setting the tone for a major class of modern control technique, what I have denominated Harm-Based Liability Rules (HBLRs).[16] HBLRs are harm-based in two senses. First, harm is the triggering mechanism: the law stays its hand until the injury has occurred. Second, the measure of recovery is calibrated to the amount of harm provable.

As a general matter, the HBLR approach has the virtue that those subject to the law—the chemical manufacturer in the more typical case—are deterred by knowledge that if they cause harm, they will have to pay for it. How and to what extent they avoid the harm is left to the managers' expertise. In ordinary domestic application, the industrial enterprises are presumed to have the motivation and competence to avoid adverse legal judgments, just as laissez-faire presumes them most capable of avoiding losses inflicted by the market. The law informs the managers that they will have to bear (internalize) the social costs of any environmental damages they cause. Firms that fail to shape up will lose lawsuits, suffer competitive disadvantage, and, rightly, decline.

Such harm-based liability is in fact the approach of several international treaties. Under the 1972 Convention on International Liability for Damage Caused by Space Objects, any nation whose satellite tumbles into another sovereign's territory is strictly liable for the damage caused. The virtues of the ex post approach here are obvious. It would be nice to insist upon precautionary rules for design and launch of satellites, but as the world now stands, nations are too secretive to allow the access such a preventive approach would require.

While there is much to be said for this ex post regulation, which in fact continues to play a large role in controlling domestic industry, it has drawbacks that international circumstances only magnify. The first, obviously, is that the law steps in only in reaction to events that have already taken their toll; some modern toxic tragedies are too serious to gamble that managers will calculate ways to avoid them, if only their enterprise is confronted with the prospect of large enough judgments down the road. The pressures of the short term (both in profit-making and in politics) and managerial disorder may make leaders systematically underappreciate the true risks.

Not only are damage suits imperfect as deterrers, in many circumstances they are not even very successful at securing compensation for victims. To institute lawsuits, victims have to know that they have been victimized. But many serious, widespread, and insidious injuries, such as workplace-triggered cancers and asbestosis, can have long and covert latency periods. Even if the injury is diagnosed, it is often frustrating—and always costly—for the plaintiff to prove which of many propagators of the dangerous substance was responsible. And even if the suit is successful, the defendant may be judgment-proof or have declared bankruptcy. The point is, the threat of having to pay a damage award is no assurance that the firm's social costs will really be internalized.

Moreover, we have already seen (chapter 3) that the analogous situation in the international arena is even worse. Nation-states (and their officers) enjoy various doctrines of immunity and "act of state" that any giant oil company would envy.

169

Penalties

At least in the domestic scene, one response to the slack left by the HBLRs has been to increase reliance on civil and criminal penalties. The idea is to confront the potential wrongdoer with a legal judgment—the "bill" for its misconduct—that is not limited to the amount of provable harm in the case for which it got caught. Penalties have also been favored in instances where the misconduct, such as pouring toxics down a storm drain, is so offensive to public morals that it seems inappropriate to leave the actor free to weigh the profits of the wrong against the damages that may be proven. The penalty is added onto the level of ordinary damages, and often enforcement is transferred from private to public hands as a means of reinforcing the message that such conduct is not just economically inefficient; it is socially intolerable.

Maybe some day such devices will be transferable to the international arena. But as we have seen, a world order that throws up a number of roadblocks to the imposition of ordinary damage claims against nations is not on the verge of empowering any international body to invoke punitive damages or fines.*

Standards

To put HBLRs, penalties, and standards in perspective, penalties differ from the HBLRs in detaching the level of sanction from the level of harm actually caused: $1,000 damage to the environment can result in a $100,000 fine. A standard, such as the requirement that asbestos be eliminated from public buildings, detaches the law from a prior harm-causing occurrence. Standards are legal requirements

* Remember that this applies to awards levied against nation-states: awards against multinationals are not so limited. Moreover, victims in Nation B of transboundary pollution emanating from Nation A can often sue the companies responsible in their own courts or, if need be, travel across the border (meaning, hire lawyers in A) to sue the companies in A's courts. See chapter 3. And while there is no standing international criminal court, some international treaties oblige covenanting nations to enact laws that may punish violators of international obligations. See chapter 5.

imposed on actors in order to prevent or minimize damage before it occurs. In effect, standards withdraw from the actor the discretion to determine the best harm-avoidance techniques "on its own," subject only to the prospect of fines and punitive damage awards should it make the wrong choices. Instead, public agencies require regulated facilities to adopt certain preventive measures as a condition of doing business.

Generally speaking, standards fall into three classes: factor, output, and bureaucratic.

Factor standards constrain the selection of input. For example, one way to safeguard air quality is to issue regulations that require utilities to install in their smokestacks a certain quality of scrubbing equipment, or to use coal whose sulfur content does not exceed some specified level per ton.

Output standards allow the managers to retain discretion over the various factors that go *into* the production process so long as they do not exceed specified limits on what comes *out*. To continue with the air quality illustration, under an output standard the regulated firms are allowed to use whatever grade of coal they want to buy and run their plants with whatever technology they prefer, so long as the output through the chimneys into the outside world do not surpass x pounds of restricted material per day. Factor standards are in some circumstances easier to monitor; for example, there are so many points of entry of GHGs that the control might have to take the form of constraints on the *use* of factors, for example, carbon, which closely correlate, for reasons of chemistry, with the undesired output irrespective of technology. But output standards generally allow more flexibility than the factor standards, which impose uniformity across an entire industry insensitive to differences among firms in terms of their companion technology, and the like. The output standards are also less likely to stultify innovative ways to produce the desired result.

Bureaucratic standards encroach on managerial discretion over organizational and bureaucratic variables. For example, if a nuclear power plant experiences certain failures or defects, the data to be collected and the pathway for that information are not left to the company to decide in its sole discretion. Under regulations of the Nuclear Regulatory Commission (NRC), certain information has to be gathered, and passed upward to officers and directors of the corporation (to deprive them of subsequent "deniability" should there be an accident). The officers, in turn, are required to pass the information along to the NRC. Procedures under the Pure Food and Drug Act impose similar pressures on pharmaceutical manufacturers to assure the adoption of appropriate internal information procedures.

Just as the criticism of the ex post remedies is that they will underdeter—allow too much wrongdoing—the primary criticism of standards is that they will overdeter—make industry put $1 million into preventing an accident whose expected cost is only $1,000 (a common criticism of the Occupational Safety and Health Administration [OSHA] regulations in the United States). What makes the charges of excess caution credible are the natural biases of rule-making agencies. An agency such as the Food and Drug Administration falls under considerable fire if it approves a drug that unexpectedly turns out to kill ten people; on the other hand, the same agency gets little credit with the public or its funding sources if by expediting approval of a new drug it prolongs or saves one thousand lives. This is a criticism of drug regulation that is frequently raised today in the AIDS context. Such imbalances are endemic among administrators. Local fire and safety codes are designed to get the code drafters off the hook no matter how extravagant the expense and unrealistic the probabilities of the peril they are protecting the public—and themselves—against.

In assessing domestic experience, I have argued[17] that standards ought to be favored over HBLRs when there is some strong combination of the following factors:

1. There are features of the anticipated hazard that render after-the-fact strategies unacceptable, even to the point of warranting some "premium" in precautionary expenditures. Such features might include a deep societal aversion to the type of harm (lethal levels of nuclear radiation, for example), the apparent inadequacy of monetary com-

pensation to make victims of the peril really "whole," and complex problems of after-the-fact litigation and proof.

2. The government's access to the relevant data regarding risk and risk-reduction techniques makes it as well informed as the industry regulated.

3. The enterprises affected are relevantly similar, or we can otherwise avoid the trap of our invoking sweeping ex ante rules that yield less benefit in controlling "bad" companies than the costs imposed by straitjacketing their innovative and generally law-compliant competitors.

4. There is a strong relationship between the variable the standard affects and the outcome to be avoided (in the sense in which there is a strong relationship between faulty auto brakes and auto accidents).

When attention turns to the international context, it is easy to see how these considerations support some increased reliance on a standards approach.[18] Clearly, the first consideration, avoidance of virtually unremediable tragedies, support the standard-setting efforts of the International Atomic Energy Agency (IAEA). Sometimes international standards will offer advantages that derive from cost considerations. It is a lot easier to regulate fishing by policing gear standards (fashioned in terms of seine size or net type) than it is to punish nations that exceed an allowable catch tonnage.

But even aside from clear opportunities to apply standards in specific instances, there is the interesting general question, whether there is not more warrant for applying standards in the international context than in domestic arenas. In my view, many considerations point toward a relatively more liberal use of standards internationally.

Most important, the relative reluctance to impose monetary judgments internationally—the unavailability of international fines and punitive judgments in particular—makes a very strong argument for switching some weight of regulation to standards. Put otherwise, when we turn to the world order, the increased risk of underdeterrence (too much harmful behavior because wrongdoers do not face the threat of the more severe ex post measures) means we may have

173

to accept a higher risk of overdeterrence (preventing legitimate conduct, the flaw of the ex ante regulations). A lean toward ex ante measures does not run against the grain of international practice. While nations are terribly cautious about exposing themselves to the possibility of ex post remedies, they appear readier to accept ex ante restrictions, even those that tie their hands internally—for example, the ozone protocols' factor constraints, which force countries to switch their industries away from CFCs at great cost. This contrast is not inexplicable. The costs of the ex ante measures, while never certain in advance, are at least reckonable enough to enter into the budget process—something that cannot be said for ordinary legal liabilities.

Finally, in regard to the concern that many regulators overprotect to stave off public criticism of their performance, there is reason to doubt that international standards-setters will incline in that direction. If only because the power of international agencies over their regulees is less secure, it is more likely that proposed tough mandatory standards will be watered down in the course of agency deliberations.[19] And even if the ecostandards that emerge from international agencies are technically nonbinding or contain easily invocable provisions for nations to opt out of them ("nonacceptance"), they can exercise considerable influence on behavior; that has been the case with the nonmandatory labor standards adopted by the International Labor Organization (ILO) and the maritime safety regulations recommended by the International Maritime Convention Organization (IMCO).[20]

THE NEW GENERATION OF LEGISLATION

Recent years have witnessed a new generation of special environmental legislation worldwide. In the United States, where law reform has been the most extensive and varied, the "new" generation goes back at least to the Clean Air Act of 1963 and includes the National Environmental Policy Act of 1969 (NEPA), the Clean Water Acts

(1977, 1982), the Toxic Substances Control Act (1976), and the Comprehensive Environmental Response, Compensation and Liability Act of 1980 (CERCLA), as amended by the Superfund Amendments and Reauthorization Act of 1984 (SARA). A full list would be numbing; our purpose is not to tabulate statutes but simply to flush out the range and vocabulary of underlying general strategies—particularly novel tactics—that might be available for international application.

Informational Approaches

The more recent laws give increasing emphasis to the production and dissemination of information. We have already discussed the environmental impact statement (EIS). A related technique is mandatory premarket testing of potentially dangerous products, a tactic adopted by the Toxic Substances Control Act. Provisions requiring warnings to consumers of suspected danger have become increasingly widespread. Merely developing a registry of dangerous substances can be a successful informational approach. On the international scene, UNEP is already working toward this end in its International Register of Potentially Toxic Chemicals, and there is a movement within the European Community (EC)—so far without success—to set up a communitywide registry for pesticides. The proponents of the EC registry are seeking to link it to tough, common standards throughout the Community; however, an approval of a pesticide by one country would provide a "passport" to all other EC states.[21]

One final informational technique deserves mention: normative pronouncements that are not so much about product characteristics, or even environmental impacts, as about shared feelings and commitments. Domestically and across the world, a lot of energy is being poured into conferences that seem to accomplish little more than to declare agreement that things are going badly and need improving. It seems like a waste. Yet such declarations are a way for the delegates to signal a willingness to participate in a common endeavor—at the least, not to conceal preferences and free ride.

Technology Forcing

Increasingly, too, lawmakers are relying on the forcing of technology. Technology forcing resembles technology fixes, such as mandatory seine net sizes. But a technology *fix* simply takes a technology that is available at the time of the regulation and makes it mandatory across an industry. *Forcing*, by contrast, involves mandating targets (ordinarily output standards) that are not achievable at the time of the rule but that the industry has to meet before a certain date under threat of penalties. Most Americans are familiar with the concept through automobile standards cast in terms of smog emissions and gasoline mileage. The firms subject to such rules invariably insist that the standards are preposterously unrealistic, destined to shut the nation's factories, and throw tens of thousands of workers on relief. The lawmakers cross their fingers, trusting that the law is one of those necessities that mothers invention. The lawmakers are usually right. Industry finds a way.

René Dubos reminds us that this is a ritual that goes back to 1863 at least, when the British Alkali Act required factories to cut back hydrogen chloride emissions an ambitious 95 percent. Almost immediately technology responded with ways that not only met the target but enabled the factories to capture what had been "waste" in the form of commercially marketable chlorine gas.[22] The same thing happened when mandatory pasteurization of milk was first proposed. The milk lobby fought it tooth and cud, because it required new equipment (and was unfamiliar). What they had not accounted for was that pasteurization, by retarding spoilage, had the consequence of increasing shelf life and requiring fewer deliveries, so that both the companies and the public ultimately benefited.[23]

One of the virtues of technological approaches (in the form of fixes or of forcing) is that it is often easier to vary technology, through a mandated standard, than it is to change human habits. Three-fourths of the litter in the Mediterranean consists of plastics. We can try to train people how to discard plastic refuse, even to fine them for littering. But any real victory will ultimately require going right to the source, the plastics manufacturers, and forcing them to develop and exploit biodegradable material.

Permitting

In the spectrum of preventive strategies, permitting is an important ally of standards, partly because the procedure provides a focal occasion for the public and outside experts to intervene before things have moved too far. While there is no international environmental agency with the licensing power of a domestic water or air quality district, there are a number of variants on the international stage. The ocean-dumping conventions include permitting procedures, although their administration is largely entrusted to each participating nation, with the consequence that enforcement is generally regarded as inadequate.[24] The Nordic Convention on the Protection of the Environment is also suggestive. It provides that each contracting state, when considering (its own) permitting of potentially hazardous activities, must take into account the effects on the other parties.[25] Although permitting has yet to meet its potential in the international arena, further evolution is almost certainly inevitable.

Accounting for Harm to the Environment

The early "environmental" strategies were predominantly homocentric in motivation and remedy. That is, while they served to safeguard the environment incidentally, the basic aim was to protect the health or property of humans who were suffering through changes in the environment. No one considered the early cases "environmental" because the courts were not being called upon to protect the environment as such. The system functioned in a way that protected the ecosystem of a polluted stream, but only when some human saw fit to complain, and even then the stream benefited as to the extent that its degradation could be "cashed out" as an economic loss to the complaining human. What went uncalculated in court were all those

177

features of nature that the market did not put a price on: the non-commercial waterfowl, the spiders and reeds and other lower life. In fact, even commercially valuable fish and animals might go unaccounted for on the view that until someone had brought them under control, "captured them," they were no one's property; and until they had become some owner's property, there was no one with legal standing to complain about their fate.

The new generation of laws has shifted emphasis away from pollution control to embrace broader conservationist, even nonhomocentric goals. The Endangered Species Act (ESA) and Marine Mammal Protection Act (MMPA) protect animals such as sea otters and threatened species even where it is clearly not cost-beneficial to do so. Indeed, in one of the most celebrated law suits under the ESA, legal actions on behalf of a small rare fish, the snail darter, derailed construction of a $100 million dam the Tennessee Valley Authority wanted to construct along the Little Tennessee River.[26]

The same solicitude for nonhuman nature has appeared in the new generation of laws that empower units of the federal or state governments to sue polluters as trustees for the environment, and to recover and apply the costs of restoring the ecosystem even if the costs of repair exceed the market-measured costs of the injury, that is, the reduced consumption and use value to humans. In effect, these new rules adopt the harm-based liability rule approach, discussed above, but clone onto them an expansive, not quite so homocentric notion of "harm": gradually harm to the environment *itself* has come to be pleadible in court.

Institutional Reform

Some of the most significant efforts aim at the reform of institutions, rather than at the manipulation of actors and their activities through conventional legal devices such as lawsuits. There is widespread agreement that "the most challenging problem for agricultural policy is to devise institutional mechanisms that will reward individual farmers for valuing [the precious resources they use] at their true social worth."[27] The same is true of fish and fisheries policies. It has been said that the greatest environmental disaster in Alaska history has not been the wreck of the *Exxon Valdez*, but the mismanagement and waste of the state's bountiful marine resources.[28] Around the world, what fisheries desperately need are politically acceptable and practically enforceable quota systems.[29]

We already mentioned that reforms in the land laws were reportedly instrumental in slowing deforestation of Brazilian jungles.[30] But successful revisions such as in Brazil are the exception.[31] In the United States, even at a time when lumber demand is weak, and many people are advocating subsidized tree planting as a means of increasing CO_2 uptake, the Forest Service is subsidizing timber cutting. Economists who have penetrated the thicket of the Service's resourceful bookkeeping estimate that the taxpayers are losing $200 million a year to support a program that rewards them with the loss of their forests.[32] In southeastern Alaska alone, the Service's efforts to prop up the lumber industry has subsidized havocking the Tongass National Forest—one of the nation's rare rain forests and outstanding wildlife habitats—not even to make boards for U.S. houses, but to be ground into pulp for export to Japan to make synthetic fibers.[33]

> Americans pay about $200,000 per working day to subsidize clearcutting in . . . the largest temperate rain forest left on earth. A Japanese firm cuts trees that were seedlings when Magna Carta was signed, destroys habitat for bald eagles and brown bears, jeopardizes the salmon industry and undermines tourism dependent on pristine scenery. For this privilege, the industry pays approximately $2 per tree, the cost of eight first class stamps.[34]

As the *Los Angeles Times* editorialized, "It would be better to pay Alaskan loggers not to work than to pay for them to conduct a firesale razing of a magnificent rain forest."[35] Congress has bridled this madness, but refuses to pull it to a halt.[36]

Part of the difficulty in achieving rational resource management is that throughout the world, resources and environmental policies are governed by a bizarre patchwork sewn out of federal and local agencies, water boards, tax and zoning codes, subsidy patterns, and more. Reforms of any real reach involve modifying entire patches and the relations among them, which is a harder job than pulling out any single thread of law. Those who want to amend a law face the opposition of those who benefit from it; but those who seek to uproot or disturb a whole institutional practice frequently find themselves taking on entrenched public-sector bureaucracies as well: not only farmers, but a Corps of Army Engineers, a Department of Agriculture, and so on.

The Embroidery of Nongovernmental Activists into the Law-making Process

Finally, one of the most important recent developments has been to provide space for the activities of citizens and nongovernmental organizations (NGOs). Much of the activity is simply informal in the sense that it occurs outside any government structure. Governments may hesitate to retaliate against treaty violators. But Greenpeace, aroused by Iceland's flaunting of the whaling moratorium, mobilized consumers across the world to bring pressure to bear on distributors of Icelandic fish.[37]

In many other ways, NGOs have operated as a source of ideas and as a conscience, gathering and transmitting information, prodding legislative bodies to higher levels of sensitivity. The credit for getting developmental banks to prepare EISs goes almost entirely to the sustained efforts of the NGOs.[38]

But perhaps the most interesting development has been to recognize and carve out a space for environmental NGOs within formal systems. It is not just that they are winning "observer" status at more and more international conferences (a dubious perk for anyone who has a low tolerance for windy orations). In many circumstances, particularly in Europe, environmental groups that can demonstrate expertise in a subject area are granted standing even where ordinary citizens could not institute a "citizen's suit."[39] This is in line with my proposal (chapter 4) that NGOs might play a special guardianship role in the protection of the global commons areas. And the Economic and Social Council of the United Nations (ECOSOC) has not only granted NGOs formal consultative status, but has provided in some detail how NGOs may participate in ECOSOC debates and decisions.[40] At the 1992 Earth Summit, the Commission on Sustainable Development clearly indicated further reliance on NGOs "to ensure the effective follow-up of the Conference."[41] Considering the increasing willingness of citizens in the present climate to support NGOs and the special role the groups can play by operating outside electoral politics, no one doubts that their contribution to protecting the global environment is destined to expand.

Taking Out Calamity Insurance

Just as lawmakers and regulators are uncertain about the impact of each of the control mechanisms they deal with, scientists and economists are unsure and at odds about their forecasts. The scenarios that raise the most concern involve uncertainties and risks on a wide-swinging, high-stakes scale.* The future may not turn out so bad. On the other hand, it could conceivably hold outright calamities in store: a nonlinear change in climate variables, a destruction of the ozone shield, a breakdown in the ocean food chain, or something even worse we haven't thought of yet. When faced with the prospect of such wide swings of fortune in more familiar contexts, our thoughts often turn toward insurance. Indeed, as scientific and economic uncertainties have accumulated, it has become common to hear references to "insurance" in the climate-change context. At the UNCED meeting in Rio, while most of the discussion of climate change focused on firm targets for CO_2 reductions, an insurance annex proposed by endangered island nations (the Alliance of Small Island States, or AOSIS) managed to slip into the Climate Change Convention in the form of one of several actions to be given "full consideration."[1]

The inclusion of "insurance" as an option is wise. But most people who have injected insurance terminology into the discussions are using the terms in a metaphorical rather than technical sense, generally to support larger expenditures on precautionary measures. For example, climatologist Stephen Schneider, in urging investments in

---

* The distinction between risks and uncertainties is not vital, but I employ risk to denote the probability that a certain well-understood state of affairs will come about. In flipping a coin, two states are possible, heads or tails. And the "risk" of tails is .5. With climate change, ozone shield thinning, and ocean pollution, we are uncertain even as to the states across which the probabilities are to be assigned.

such CO_2-mitigating techniques as energy efficiency, notes that "like an insurance policy . . . this protection does not come free. It requires a premium of a few tens of billions or perhaps hundreds of billions of dollars spent annually around the world."[2] William D. Ruckelshaus, the former EPA administrator, has observed that "insurance is the way people ordinarily deal with potentially serious contingencies, and it is appropriate here as well. . . . Current resources foregone or spent to prevent the build-up of greenhouse gases are a kind of premium."[3]

I agree that some level of investment in energy efficiency is justified. But it is not at all clear that tens or hundreds of billions of dollars is the right "premium" or even that "insurance" is the right term for what Schneider and Ruckelshaus and others have in mind.

To clarify the terminological point, let us put insurance in the broader context of risk management. Conventionally, techniques for dealing with risk are divided into sometimes overlapping categories of *avoidance, reduction, control, transfer* (including sharing), and *retention*. To illustrate, imagine someone who is considering living in a hillside area that is periodically exposed to devastating fires. The prospective purchaser can *avoid* the risk by simply forgoing the opportunity to live in the hills; *reduce* the risk by, say, cutting back brush; *control* the risk (or level of loss) by installing a sprinkler system and smoke detectors; *transfer* part or all of the uneliminated risk to another party, ordinarily through an insurance company (whose charge can be expected to reflect the insured's efforts at risk reduction and control); or *retain* some remainder—that is, leave it uncovered and live with it. Two things should be observed: (1) the rational solution is to deploy each mode of response in a manner that minimizes the sum of their costs;* and (2) only the fourth, transfer, involves "insurance" in the ordinary commercial sense of the term.

---

* To illustrate, imagine the owner of a house valued at $500,000 who faces a one-tenth percent probability of total destruction by fire in any year. If she is at all risk averse, she will be willing to pay an annual premium of somewhat over $500 for a year's full loss coverage. But suppose further that (1) by cutting back brush and installing sprinklers at an annual cost of $300, the risk of a total loss were to fall to two-tenths percent and (2) that the insurer's rating system accounted for

When we review the climate change and other future perils literature from this perspective, many of the options discussed can be
described in risk management terms. The goal of dealing with perils
is, essentially, to achieve the least cost combination of the risk-man-
agement categories. Strategies aimed at preventing the accumulation
of GHGs can be categorized as risk-avoiding or reducing. So, too,
can some of the countermeasures under consideration that would
remove CO_2 (aforestation, nourishing oceanic phytoplankton colonies) or nullify climatic effects of the buildup (such as injecting aerosol reflectors into the atmosphere). Seawalls and improved water
systems can be viewed as risk-controlling: they accept the increased
climatic hazards (more storm surges, more droughts), but aim to
minimize the economic losses they cause.

Note, however, that none of these measures really involves "insurance" in the risk-spreading sense. Those who advocate greenhouse
"insurance" typically appear to be suggesting that we can practically
avoid or significantly *reduce* the risks of perils associated with climate
change by accepting the costs of stabilizing atmospheric concentrations. That is to use "insure" in the loose sense in which one might
say he has "insured" a house against theft by paying to install an
alarm system, meaning he has made a robbery less likely to occur.
But the use of insurance in this sense does no more than invite a
reexamination of the costs/benefits question. Once more: an ounce
of prevention may be worth a pound of cure; but just as clearly an
ounce of cure does not merit a pound of prevention.

Those who justify prevention expenditures as "insurance" may be
intuiting a point about risk aversion, or perhaps even about the utility
of money in certain states. Suppose that the mathematically expected
total loss of a house from fire is $500 a year, that complete insurance
would cost $510, and that to practically eliminate the risk by brush
cutting and sprinkler installation would cost $1,000 a year. While
$510 is the least-cost option, expending the $1,000 for prevention is

---

those precautions by reducing her premium to $110. The sum of prevention plus
insurance thereby drops to $410, and the owner would be expected to prefer that
combination.

not irrational. It depends upon the homeowner's taste, not just for risk, but for the sort of risk involved, and how she would value the proceeds of a policy in the circumstances she would then find herself in.* The owner may be so averse to an economic loss that occurs *that way*—through a horrible life-disrupting conflagration that sweeps away irreplaceable belongings of sentimental value—and her needs in the devastated state would be so altered that she is prepared to pay $1,000 for, let us call it, "'insuring' against the occurrence."[4]

In the same vein, the climate-change discussion might well focus on whether the peoples of the earth have (or, if fully informed, would have) a collective aversion to some of the highly unlikely but also highly unlikable possibilities that global warming could bring. To the extent they do, there would be warrant for preventive expenditures beyond the amounts justified by more conventional assumptions of risk-neutrality.

That appeal strikes me as theoretically reasonable—although, practically, any variations in policy that might be implied from defensible attitudes toward risk may well be swamped by the implications of defensible discount rates and, indeed, of how one resolves the philosophical conundrums of valuing the welfare of future generations.**

---

* Once considerations of state-dependent utility are introduced, the analysis becomes considerably more complex. Suppose (in normal times) I am considering taking out an insurance policy against the loss of my home to flood. It is not too difficult to judge what lump-sum payment arranged with the insurer would make me equally well off as I am now. But suppose that the insured event will occur in a world truly savaged by climate change. As I consider insurance, I have to wonder what utility I would get from the insurance proceeds in such circumstances: would there be—however much money a surviving solvent insurance company were to place in my hand—adequate food? water? medicine? Perhaps I would shift from true insurance to risk prevention—just as Schneider and Ruckelshaus intuit. On the other hand, some extreme post-deluge states of the world might be regarded as so ghastly, and our choice set so limited, that we would react with neither prevention nor insurance, but with increases in present consumption.

** Suppose a deity offered us a gamble: instead of whatever the future will bring if she does not intervene, she can guarantee the following odds and payoffs: (1), with a 51 percent probability, in 2100 there will be twice as many people on earth as

Moreover, even if we can provide a coherent account of generation-skipping risk aversion, it is not immediately apparent why any allowance for risk aversion is better accounted for via preventive measures such as emissions reductions than via, say, risk-spreading techniques such as true insurance. In fact, insurance allows each individual (or nation) to tailor its protection according to its own relative tastes for risk, consumption, savings, and the value it places on its descendants' lives and well-being. That degree of individualization is hard to arrange if protection takes the form of a joint undertaking, for example, a collective multinational decision to stimulate ocean algae with iron ferrules, or to mirror the seas with polystyrene.

Most important, to defend present expenditures of hundreds of billions of dollars, it is not enough to point to the need to make some accounting for risk preferences and declare that commitment "the social risk premium." Schneider opines that immediate avoidance measures "would merit [as a sort of insurance premium] a few tens of billions or perhaps hundreds of billions of dollars spent annually around the world." But would it? Nordhaus, recall, estimates that $300 billion per year would get us a 60 percent reduction in emissions worldwide; but he also estimates the net benefits of the 60 percent reduction as only $55 billion.[5] If risk aversion is to be the warrant for an apparent imbalance of dollars spent and value received, we ought to frankly explore the issues: Does talk about the whole world's risk aversion make sense, and if so, what are we prepared to pay in risk premium? What are the implicit linked judgments—about intergenerational discount rates and the relative marginal utility of wealth—to us and to our descendants? Is there a prospect of a superior alternative through some combination of outlays for risk avoidance, risk control, and insurance?

---

today, each equally happy on average as today's population, or (2), with 49 percent probability, that a cataclysm will leave only 1,000 miserable survivors. Would we accept the gamble? If not, what would we pay the deity to change the odds to 60–40?

THE APPLICATION OF CONVENTIONAL INSURANCE
TO CLIMATE CHANGE

In many ways, the application of insurance in the commercially conventional, risk-spreading sense appears quite suited to many of the perils of climate change. Global warming, if it develops, is expected to produce an uneven distribution of presently unidentifiable winners and losers, with many commentators refusing to dismiss the possibility that even with no market-forcing reductions, local gains will, on net, dominate local losses.[6]

We know, too, that whatever the possibility of overall gains, it does not make the prospect of local losses less calamitous to those who will suffer them. For the small island states, countries such as the Maldives, Vanuatu, and Kiribati, which are imperiled by a rise of only a few meters in sea levels,[7] it is not only a question of lost property and of transporting populations to some new locale, but the prospective losses include what we might call "diaspora" costs: the untying of a people from the links to their land. And in the aggregate, such widespread "local" disruptions can have serious repercussions on the global community as a whole.

If that is one's picture, then some mechanism for risk transfer and spreading, *in efficient combination with avoidance and control measures*, should play a prominent (although of course not exclusive or necessarily even predominant) role in a climate-change policy. The notion is not that anyone would write policies naming "climate change" as a peril. It is not a massive unitary disaster—the destruction of civilization, a biblical flood—that would be insured against. To a large extent, what global warming portends is the gradual intensification, or shifting location, of perils many if not most of which are insurable in present markets: windstorms, inland flooding, freezing, sea surges, health, and crop losses, and the like. Such insuring is already a substantial business. In the 1970–89 period insurers worldwide expended U.S. $46 billion in covering insured natural peril losses.[8] Indeed, European insurers (recognizing that when the academic talk about global pollution trading has blown over, some role for them is a foregone conclusion) are already examining how much they may be exposed to claims that greenhousing may add in the form of more intense and frequent weather losses.

But the issue for us, the larger society, is not just one of business, of how insurers would fare in a changing climate. From the global public's perspective, the question raised is part of a larger subject: the capacity of various private markets to smooth the transition to a warmer world—for example, the capacity of commodity prices to signal appropriate adjustments in the supply and mix of commodities.[9] As a policy instrument, insurance not only has the capacity to soften (by spreading) the financial blow of any losses that fall, but it may send signals calculated to minimize aggregate losses over time. To illustrate: as certain crops, or life in certain areas, became more vulnerable to climate threats, insurers would be expected gradually to raise premiums for the relevant classes of risk. People who persisted in growing increasingly imperiled crops or living in increasingly imperiled areas would have to pay for it. Assuming that the insurance companies mobilize the expertise and the premium payers seek to minimize their premiums, global investment should be at least nudged in the right direction—away from the most vulnerable acts and activities.

In fact, we may usefully think of taxes on emissions as dampening social costs by sensitizing risk generators, while premiums collected through first-party property insurance operate in the same direction by sensitizing risk-bearers. The more universally we identify risk-generating activities and the more bilateral a view we adopt of the harm—that is, the less we regard coal burning and rice farming as intentional torts like hitting one's neighbor with a tire iron—the more the choice between the two approaches is open to efficiency rather than to other moral considerations.*

* For example, we should examine which device, taxes or insurance, is more likely to minimize social loss through administrative error. My guess is that insurance categories have been easier to modify beneficially in the face of experience—although, if so, it may tell us more about the distinction between politics and markets than that between taxes and insurance.

Moreover, taxes and insurance are not inconsistent options. One alternative is to impose a modest fee on the dischargers that is safely within the consensus range of the estimated externality, and to use the proceeds to make up for shortfalls in an insurance plan that is chiefly funded from private (or national) premiums, but that also recognizes the inability of some insureds to pay their way. Such a scheme could be structured to combine some elements of a third-party insurance system (in which potential wrongdoers arrange compensation for losses that the victims may shift onto the wrongdoers through legal actions)[10] with elements of a first-party system (in which potential victims arrange for compensation directly from the insurance carriers).

The private insurance sector has a lot to offer. To begin with, the industry commands considerable sophistication both in assessing risks and in assembling risk-shifting portfolios. For example, the network of insurers and reinsurers would be expected to hedge. That is to say, if, as time passed, warming-associated losses became increasingly probable, the companies thereby exposed to higher claims would shift investments into industries that would be expected to realize gains from warming, such as manufacturers of air-conditioning equipment and builders of sea walls. The social desirability of the resulting risk distribution, as well as the influence of the investment classifications, is evident.

Limits on Private Market Participation

On the other hand, there are distinct limits, from both a commercial and a public point of view, to what we should reasonably expect private markets, unreinforced by government participation, to provide.

Even if global warming does not inundate the industry with a sudden flood of claims, questions have been raised about the capacity of private-sector insurers to survive an increasing barrage of "natural disaster" losses. Catastrophe insurers, worldwide, are already under considerable financial pressure.[11] The wave of insured losses in the 1980s may have reflected nothing more than an aberrational bad spell of weather in northern Europe. But industry analysts worry that, even without allowing for more climatic disturbances, the firms face spiraling payouts just from projected increases in population density and property values.[12] Then, too, if the world gradually becomes imperiled by climate change, insurers will be vulnerable to adverse selection—of being left to insure only the worst risks while the safest insureds, balking at the escalating premiums, withdraw from coverage.

On the other hand, the insurance industry's own perils should not be exaggerated. Gauging the net impact of global warming on insurers, as on anyone else, is complicated. One expects that the higher claims faced from flooding in some areas would be offset by reduced claims for floods, freezing, and crop loss in others—exactly the sort of risk spreading we hope to achieve from globe-spanning insurance. Moreover, the industry has any number of ways to protect itself financially. It can raise premiums, withdraw from certain areas of coverage, and write policies with more guarded exclusions, limits of liability, cancelation clauses, and so on.[13]

Inability to Pay Premiums

The second concern has already been signaled. Continuing to focus on first-party plans, the viability of private coverage turns on the ability of risk bearers to come up with the required premiums. As risks increase, insurers will do their best to single out classes of high-risk properties, for example, coastal developments, for higher premiums. Poor countries cannot realistically be expected to pay their actuarially fair share of the premiums, which is serious because a larger share of the LDCs' economies is typically invested in climate-sensitive primary activities such as farming. Further, enjoying fewer economic and institutional resources, they are apt to be less resilient should climate variables turn unfavorable. A developed country can effectively self-insure, absorbing multibillion dollar losses from natural catastrophes that would wound an LDC mortally. Indeed, a poor country may leave more wealth exposed, even in absolute terms, than a rich country as a result of not having the same ability to protect its assets with, for example, dikes and flood control systems.

Hence, there is a strong argument for some international arrangement to account for the premium hardship, whether by subsidizing premiums or otherwise (see below).

Limits of Coverage

Even if the financial health of the industry were unquestionable, and even if there were assurance that premiums would be forthcoming from or provided on behalf of poor nations, it is far from evident that private insurance would meet widespread concerns over climate change.

To begin with, not all the perils posed by greenhouse warming are insurable under present industry practices. One major class of private losses falls outside commercial coverage entirely. Catastrophe policies are available for sudden, periodic but unusual happenings: a house swept away by flood, a crop lost to late freeze. However, many of the greenhouse-associated losses are apt to arise gradually and almost imperceptibly: through erosion of beach front or decline in soil moisture. In those circumstances, property values drop, but insurers resist covering a decline in value as a form of property "loss."

Without contesting the soundness of this position—complicated assumptions about causal proof and moral hazard are involved—its persistence would mean that, lacking government intervention, there would be no insurance against the risk of gradual declines in property value, one of the major categories of greenhouse concern.

Second, private coverage is unavailable for the full amount of losses, particularly in regard to large-scale perils. Commercial policies typically impose limits on claims (per incident, per policy, etc.), recite certain exclusions by class of event, and leave the insured with some retention (deductibles). We have already glimpsed the commercial justifications for these policies, justifications that seem to assure their perpetuation. The industry has to protect its own well-being. Besides, mandating full coverage would exacerbate moral hazard: the concern that insureds, their fear of loss blunted by the coverage, will fail to take efficient risk-control measures.

Third, insurance protects owners against their private losses. But the losses that fall under the shadow of greenhouse warming include injuries to goods that are public or collective. There are a number of these: the imperiling of shipping lanes with icebergs, impairment of biodiversity, the marring of commons areas such as the Antarctic, the submerging of an area that the world considers part of the global cultural heritage, and heightened geopolitical tensions. These are among the very solemn losses that lie outside what private insurance as we know it, or practically can conceive it, will cover.[14]

Modes of Government Participation

We have seen several factors that favor private coverage. But we have also noted a number of factors that appear to make some public role almost inevitable. From the industry's point of view, the magnitude of the exposure, the time frame, and the industry's present financial uncertainties can all be expected to restrict participation. From the public's perspective, doubts about the capacity of the private sector to manage a problem of this magnitude, and the "gaps" in protecting, for example, collective goods, combined with widespread distrust of private markets abroad, would operate in the same direction. Then, too, I have been implicitly assuming that the insurance would run to private parties; but it is easy to imagine nations desiring to establish themselves as beneficiaries (as apparently do the island states, discussed above), an arrangement for which some sort of public international law regime would probably be required.

The question thus arises of government and international participation. Government-supported private insurance in selected markets, and, indeed, specialized government insurance, is not unknown. The United States alone presents a rich variety of models. For example, there is government-backed crop insurance; the Overseas Private Investment Corporation, which insures U.S. corporations operating abroad against the risk of being nationalized; and plans for compensation in the event of nuclear accidents, which limit coverage but backstop private market shortfalls.

Government-Supported Private Market Insurance

One option is for the international community to reinforce the private markets. In this role it could subsidize premium payments, particularly on behalf of high-risk LDCs; make up politically perceived "gaps" in commercially available coverage arising from exclusions of peril, retention requirements, or losses in excess of insurable limits; or generally reinsure on better-than-market terms.

Presumably the required funds could be raised via fees based upon some mixture of the following: each nation's expected benefits; indices of national ability to pay, such as used by the United Nations in determining assessments; and national levels of GHG emissions.

Would government subsidy in this area be wise? One commentator has objected that "it would be a mistake to counteract [decreases in privately available insurance] through a program of public reinsurance of such policies, for much the same reasons that it is a mistake to encourage people to settle in flood plains by providing subsidized flood insurance." [15]

The point is the right one to raise. Why should the global community choose to override the market and subsidize coverage for a nation that continued to develop coastal properties, or switch crops, in the face of changing levels of peril over decades?

The answer, however, is far from self-evident. First, governments commonly encourage some of their population to take greater risks than markets would underwrite. Subsidizing some of the population to settle on a flood plain may be a perfectly sensible government policy.

Second, Kenneth Abraham calls attention to the normative complexity of deciding the factors for which an insured can be held morally responsible. The problems are too vast to engage here, but, to paraphrase Abraham, can poor populations established along coasts

193

of low-lying developing countries be held responsible for where they live?[16] Can we (morally) meet their appeals for security with talk about moral hazard and their duty to mitigate "by moving somewhere safe"? Some degree of global subsidy might carry a moral imperative.

Moreover, if the global community decides that moderating greenhouse threats will require, or justify, a subsidy of some sort, we have to consider the alternatives to doing so via insurance. Is protection via subsidization of insurance more or less affordable than protecting imperiled nations via other devices? From the developed world's perspective, insurance might well be cheaper than comparable protection achieved through a mandatory pro rata reduction in GHG emissions or a marketable permit system that would transfer massive amounts of wealth to the LDCs in a lump sum up front.

A Fuller Government Role

At the other extreme, the community of nations might prefer to provide some of the insurance outside the industry as a supplement to what private markets will cover. There are several options. One approach is the *uninsured risks model*, in which the emphasis is on private property insurance, but with the government stepping in to cover a range of rejected applications along the lines of a government-operated uninsured motorists program. Another tactic would be to model some greenhouse insurance more along the lines of *health insurance* than of property insurance. Property insurance provides the insured with a lump sum cash payment contingent upon what the industry regards as a "loss"—the house swept away, the crops destroyed. (The amount is ordinarily the lower of replacement costs or market value at the time of loss.) Gradual reductions in value are not covered. By contrast, health policies reimburse the insured for expenses actually incurred in the course of treating an illness.[17]

Theoretically, one cannot determine which model ought to be preferred without entering into a morass of assumptions about such factors as the state-dependent utility function of a "sick" nation, or of a healthy nation in a "sick" world. But it is easy to imagine scenarios in which a nation would prefer a policy that contributed toward the cost of building defensive resources over one that promised to compensate property losses after the fact. Think of an island nation concerned about inundation. A health-modeled policy would enable it to treat the early signs of sea rise the way one treats the early signs of disease: one intervenes by medicating (building sea walls) before the situation becomes worse or fatal. Whether and to what extent such an arrangement would be ideal from the subsidizing world's perspective is indeterminate in the abstract. One can think of circumstances in which environmental "medication" to maintain the status quo—at the extreme, the analogue to intensive-care life support—could prove to be a more expensive commitment than simply "writing off" the property as a total casualty loss.

Another approach would be to supplement private markets with an *emitter-funded model* designed along the lines of the U.S. Superfund, created under CERCLA. Under that model, the risk generators, the GHG emitters,* would pay into the fund a fee based upon emissions. The resources of the fund would then be available to remedy or mitigate the effects of climate change. But it would also provide the deterrent component of third-party insurance. That is, while the GHG levy would presumably be set lower than the rate of tax required to stabilize concentration, it would nonetheless create some incentive for nations to wean themselves from GHG reliance.

Indeed, for some classes of loss the fund might be authorized, like the Superfund, to subrogate against harm causers according to some joint contribution formula based on emissions in excess of an established baseline. In other words, a party that was damaged $1 million could draw its reimbursement directly from the fund. The fund would then acquire the legal rights of the damaged party to recover the $1 million from those responsible. Causal responsibility being essentially ambiguous, the polluters would simply restore the $1 mil-

---

* Or the emitters of ozone-depleting agents, etc.

lion in proportion to how much their emissions exceeded their allocations, perhaps, for example, their 1990 levels. This would simplify the legal proceedings considerably.*

Another possibility (not inconsistent with any of the others) is that, in preference to transferring substantial sums to a fund, nations might simply agree by treaty to *pooled coverage*, kicking in payments only if and when required to cover one of the parties' losses. Such an arrangement avoids a separate earmarked insurance fund and simply looks to the general growth of international assets, and the good faith of the covenantors, for assurances that payments will be made.

An international convention could also address the gap in coverage for *collective and public goods*. As indicated, perils to such things as cultural monuments and endangered species lie beyond the coverage that private firms are likely to provide. But an insurance program built upon public international cooperation could be designed to address them. Imagine, for example, a species whose continued existence might be imperiled by climate change. The species could be named the beneficiary under a special policy whose premiums would be paid out of public funds. As part of the design, a governmental or nongovernmental organization (such as UNEP or the World Wide Fund for Nature) could be designated as the species' trustee, with the understanding that any proceeds provided by the policy would be applied by the trustee to the benefit of the species (as by underwriting the creation of nurseries and sanctuaries).

Note, however, that some collective losses that greenhouse warming might cause, such as the costs of increased international tension, probably lie outside any insurance coverage, in whatever manner public and private plans might be combined. And, of course, to the extent that is so, the argument for relying on risk-avoidance and mitigation techniques strengthens proportionately.

---

* On the other hand, the experience under CERCLA thus far can hardly inspire its close reproduction at the international level. Much of the payout appears to be going to lawyers and administrators.

THE INTERNATIONAL PRECEDENTS AND PROSPECTS

International agreements expanding insurance coverage beyond that which is conventionally available have already become an accepted institution. The International Fund for the Compensation of Oil Pollution Damage compensates parties that cannot otherwise obtain full and adequate reimbursement for their injuries.[18] Another variation responds to the threat of nuclear accidents. One of the international liability agreements in the nuclear energy field creates an insurance pool among the contracting states, contributions being established by a formula based partly on each party's respective GNP, and partly on its fraction of the parties' total reactor capacity.[19]

Each of the various approaches has its own problems that negotiators would have to iron out. Under many of the options, international regulators would have to decide whether to insure against climate-driven losses that had become too chronic and long-term to insure under conventional standards—for example, the permanent loss of habitability and the uprooting of populations. The deeper we got into such areas of coverage, the more we would politicize what are presently managerial decisions such as the classification of risks, the response to moral hazard, and eligibility for benefits. Would the regulators look only to dollar loss or account for wealth indicators such as per capita GNP? If greenhousing damage should become widespread, selecting which damage sites to repair, like prioritizing clean-up sites under CERCLA, would be politically challenging, to put it mildly.

Then, too, the causes of casualty losses are not easy to pin down. In November 1991, when thousands in the Philippines drowned in the wake of tropical storm Thelma, the governor of Leyte put a thought-provoking twist on the tragedy, blaming excessive logging that stripped water-holding forest cover from the hills. "This calamity is practically 100 percent because of illegal logging," he said. "It's a man-made disaster, abetted by nature."[20] How does a liability or insurance system account for such contributions by the "victim" nation, whose own imprudent failure to manage forests and flood control projects and levees abetted the forces of nature and the roguery of GHGs? Creation of a new international insurance agency would almost certainly be required to establish and perhaps even adjudicate the required regulations.

ALTERNATIVE "INSURANCE"

Finally, we have to examine ways to "insure" against climate-driven losses in a risk-spreading sense even outside the mechanisms we conventionally identify with insurance plans. Indeed, it is wise to remind ourselves that cultures have been devising ways to buffer themselves against the vicissitudes of Nature since the start of civilization. The means they have chosen include diversification, physical storage, cultural conventions regarding sharing, and so forth. The problem is hardly "new" and is even traced out in an engaging little volume called *Bad Year Economics: Cultural Responses to Risk and Uncertainty*.[21]

Nonmonetary Insurance

Any risk can be buffered against by portfolio management. In effect, by spreading through worldwide capital markets, the portfolio investor purchases "insurance" that is not subject, like conventional insurance, to withdrawal in the face of increased peril. Obviously, this is more of an option for developed nations and their citizens than for the LDCs. But some of the greenhouse perils can be provided for by building portfolios of special "real" assets, which can offer benefit globally. For example, one of the fears is reduction of biodiversity, in particular, eradication of the yet-to-be inventoried stock of genetic plant material. What is being sacrificed is Nature's own storehouse of genetic material, which has been available in the wilds for cross-breeding, as required, with dominant agricultural stocks that fall prey, as they do, to novel pests and changing climatic conditions. Farmers can protect themselves to a degree by taking out crop insurance. But from the whole globe's perspective, the superior response has been for nations to cooperate in establishing germ banks in which they store a broad "portfolio" of genetic codes.[22] These germ banks may be a model for a whole variety of problem-specific forms of "real" insurance that the world could exploit, for example, strategic inventories of protected habitats and even (although not my favorite alternative) zoos.

Cooperative efforts to develop countermeasure technologies can be viewed similarly. They are ways in which the community of nations pools a fund of knowledge, rather than of things, no one knowing for certain at the time of commitment which nations will be forced to make the largest "claims" upon it.

A second form of nonmonetary "insurance" involves investment in institutional flexibility. Lester Lave, noting our inability to predict even the overall value of climate change, rightly recommends that we devote resources to "enhancing education and capital formation so that we could minimize the effects of adverse changes and take advantage of beneficial ones. Since carbon dioxide induced changes are likely to be only one, possibly small, source of change, there is all the more reason to devote resources to making our economic and social institutions more flexible and adaptable."[23]

Indeed, the potential "returns" on investment in institutional reform are considerable. Many perils linked to climate change are exacerbated by flaws in societal arrangements such as tax and resource-use policies that encourage abusive overuse of forests, soil, water, and fertilizers. One good way to counteract the threat of famine from regional crop loss is to remove barriers to transportation and trade, thus paving the way for rapid relief. Obviously, such reforms will be valuable even if the perils of global warming never eventuate.

Global Social Insurance

Even if we restrict our focus to monetary devices, conventional casualty and health policies are not the only ways to insure against future losses. Another quite relevant model is suggested by federal disaster relief in the United States, under which federal resources are available to assist states in preparing for and responding to major calamities.[24] There already exists a variety of international "relief" organizations geared to these tasks, including the United Nations Disaster Relief Organization (UNDRO), the World Food Organization (WFO), the International Red Cross, and the International Organization for Migration (IOM). Fortifying such agencies would be a good "insurance" policy, the more so because their aid is not specific to global warming perils but is available for other catastrophes—both known, such as earthquakes, and those we cannot imagine.

Going a step further, we could consider a worldwide safety net of general welfare benefits, not confined to compensate losses from climate change or even from natural disasters per se, but to stave off destitution from any cause. We think *their* problem will be warming; what if it turns out to be an unforeseen epidemic, or unforeseeable complications of an asteroid encounter? Considering our gross uncertainty about the future, should not some of our global insurance take relatively nonspecific forms—provide "all-perils policies," as it were? Here, too, we already have the foundations for such "insurance" in a growing network of development banks and relief agencies.

Indeed, that leads to a final point. One would think that the best way to "insure" against remote, ill-defined perils would be to assure the general financial health of the globe—a sort of (to carry out the image) global self-insurance. Perhaps the most significant international contribution would be cooperative investment in a broad and imaginatively risk-spreading portfolio of social projects: the genome project, fusion research, bioengineering, desalinization, solar power, and so on, with provisions for ready transfer of technology.

Paying the Bills: Toward a Global Commons Trust Fund

B<small>Y RECAPITULATING</small> the myriad challenges, and even raising what many will consider the fall-back alternative of insurance, it is not my intention to discourage concerted action to deal head-on with the countless problems we face. One has to appreciate, however, the breadth of scientific uncertainty and the many sources of legitimate disagreement. It is unsurprising, and should not be advertised as a mere question of purging "indifferent" politicians, that nations have a hard time agreeing on a far-reaching and highly detailed global agenda. For every perceived problem, there are large differences of opinion as to the real risk presented, the level or expenditure that is warranted, and the appropriate policy instrument that is best deployed.

But the question is not whether rational actors should take immediate measures to safeguard the global environment. We can go back and forth on whether it is worth $1 trillion annually to knock 2° C off of the global temperature in 2035, if that target is even within reach; or whether the world community should restrict certain activities or expand international criminal law to encompass more environmental crimes. However, while we are arguing questions like those, no one should doubt that considerable outlays are warranted right now for fact-finding, monitoring, and model building. At all levels—global, regional, and local—there are vital reforms to achieve in legal and social institutions. There is plenty to press for enthusiastically on present evidence: programs that would transfer technology, provide medical assistance, train in the use of pesticides, treat water, improve safety at or shut down perilous nuclear reactors, preserve ecosystems, protect species and areas of world cultural heritage, identify and "contain" radioactive and toxic waste deposits, promote energy conservation, and more. Considerable sums are required, now, just to do the scientific cataloging required while a really first-rate biodiversity treaty (hopefully one that is stronger than the convention put before Rio) is prepared.

Hence, in face of all the confusing detail and debate, we should not lose sight of the fact that there is no shortage of solidly valuable present options. What is lacking even more than the understanding and the willpower are resources. Often all that is needed are start-up costs for programs that, as with some energy conservation measures, are almost certainly cost-beneficial.

The question then is not what projects one would undertake if we had unlimited resources, but how we can fund the things that most need doing. Agenda 21 that emerged from the Rio conference easily produced a list of aims that would require $600 billion a year to meet. Maurice Strong, who presided over the conference, indicated that developing countries will need about $125 billion a year in aid to pay for the new "environmental" programs, or $70 billion more than all the financial assistance they now receive. Japan has indicated it would respond with something on the level of a billion dollars, but some of that will almost certainly be a recharacterization of existing aid programs. The United States, for its part, already provides $11 billion in foreign aid and the Bush Administration was not prepared to grant any new environmental funding beyond the $750 million promised to fund specific international environmental projects in 1993. At the end of the conference, less than $5 billion in new money had been pledged.

The estimates on the need side may well be dreamily inflated and certainly include nonenvironmental components. The Clinton administration may open its pockets more liberally. But any way one counts it, there is still destined to be a considerable shortfall between what reasonable people would agree was vital and what is likely to be forthcoming. Can that deficit be pared?

Let us examine the options.

THE GLOBAL FINANCING OPTIONS

Voluntary Contributions

To begin with, it is important to understand that most international activities are funded on a voluntary basis. Practices vary. NGOs such as Greenpeace depend largely on the largesse of individuals. Some U.N. agencies, such as UNEP with its $70 million budget, rely on contributions from nation-states. (Peacekeeping efforts, such as those in the Persian Gulf War, similarly rely, perhaps abetted by a little stronger diplomatic arm-twisting than the environment can muster.) Other organizations, such as the World Cultural Heritage Fund, draw support from both government and private sources contributing or not, as they choose. All this takes place without the least sanction for nonparticipation.

To understand the consequences of such heavy reliance on purely voluntary giving, it is useful to consider what we have learned from the thriving charitable sector in domestic societies. Unlike authorities with taxing power, managers of local charities must continuously appeal to resource providers who are under no compulsion to give. As a result, charity managers are disciplined largely by "market" forces—the supply of charitable funds—rather than, as with managers of public agencies, having the power to tax coupled with accountability to electorates. Nonetheless, evidence suggests that often (not always) the efficiency of charitable institutions and the responsiveness of service providers to resource providers and clients compares quite favorably to public-sector agencies providing comparable services.

On the other hand, because donors are found predominantly among the wealthy, the charitable sector bends somewhat toward the tastes of the wealthy—to the advantage of art museums and opera companies. Most commentators believe that public preferences would probably be better reflected were the funds presently diverted by the wealthy *around* the tax system, to be called back into tax channels and redistributed via political institutions.[1]

203

When we turn to the international stage, we can see the same forces at play. Reliance on voluntary contributions focuses efforts on the preferences of the wealthy—individuals or nations, as the case may be. If allocation of global funds were put to a popular vote worldwide, Rio left no doubt that much would shift from saving tigers and warding off climate change to other projects more pressing to the world's masses. And that tension underlies much of the current controversy regarding management of any international fund. Characterizing their contributions as "voluntary giving," the donor nations want a fund governed along the lines of, or even under the control of, the World Bank. That is because the Bank's management departs from one nation, one vote to reflect the amount of funds that a country is providing. By contrast, the LDCs want the funds distributed by a new agency (a "Green Fund") that will be governed on a one-nation, one-vote basis, irrespective of contributions. Obviously, the conflict over governance (discussed more fully below) will be, as long as it lasts, a major factor in retarding the development of a fund-collecting mechanism.

Assessments

Not all international funding is purely voluntary. The U.N. system (including many specialized agencies such as the International Atomic Energy Agency) goes a step further in levying assessments on members based upon rough measures of wealth and ability-to-pay criteria.* Although no marshall will come to the door of a nation that fails to meet its assessment, the exaction is backed up by the threat of loss of vote and interest on arrearage.[2] In practical effect, therefore, the funding for many U.N. undertakings might be viewed as the product of a rudimentary, de facto tax system already in place.

On the other hand, the fact that it is not a *real* tax system has strong ramifications. The "levies" are more modest than they would be if there were a taxing power. And the United Nations lacks the

* In practice, several of the agencies have supplemented their budgets through a number of devices, including private contributions, stamp sales, and occasionally even fees.

taxing power. The United States is supposed to pay 25 percent of the U.N.'s budget under the organization's assessment formula, but as of August 1992 it had fallen $550 million in arrears; it has never been stripped of its vote, as it might have been after two years of missed payments under U.N. rules. (The Bush administration has committed to pay the debt down, unhurriedly.) In 1984 the United States, and, in 1985, the British, withdrew from UNESCO charging that the organization was hostile to free markets, a free press, and human rights,[3] that it disregarded budget restraints, and that it suffered from plain bad management.* When the United States withdrew, a motion to bring the United States before the World Court was dropped, reportedly because it would be a waste of time and the flap might just encourage other Western nations to leave.[4]

Indeed, it is worth bearing in mind that at some level *all* participation in any international organization is voluntary. At Rio, the United States ridiculed the proposed Biodiversity Treaty on the grounds that, while in principle the developed countries were to pay the developing countries the costs of complying, the institutional structure and criteria for assessment were left utterly blank, to be filled in by subsequent action of the parties after the treaty had been signed.[5] While the Biodiversity Treaty was fairly feeble and did not deserve much enthusiasm on that ground,** it ought to be said that the open-

---

* One charge was that UNESCO spent 80 percent of its budget at its Paris headquarters.

** Why the Convention on Biological Diversity receives such widespread support among environmental groups, who ought to have been pressuring for stronger commitments, is a mystery to me. The idea of a treaty on biodiversity *sounds* good; and there is a wing that seems to believe that if the U.S. refused to sign it, it *must be* good. The Convention may be viewed most favorably as the basic framework for something better to come. But as things stand, the agreement reinforces the right of each nation to control its own resources and consequently puts signatories under very little real pressure to protect their environments. The drafters rejected language that would have proclaimed biodiversity a part of the Common Heritage of Mankind (all too threatening to sovereignty) in favor of calling it part of a "common concern." What emerged appears as much a vehicle to transfer funds from rich to poor in the name of biodiversity, as it does a bona fide effort to come to grips with what is, all must admit, a diplomatic minefield.

endedness of the funding mechanism was not as ridiculous as it was made out to be. The funding of all joint international undertakings depends, in the last analysis, on consensus, even if it is not put in so many words. If any of the major players feel they are being unfairly treated, they can withdraw and, if they are large enough, bring down the scaffolding with them.

Taxation

In the eyes of those favoring strong world government, the freedom to opt out of, even "veto" a program is the heart of the problem, and has produced arguments from time to time for investing the United Nations with a more conventional power to tax.[6] Undeniably, doing so would produce a more dependable flow of funds from year to year, to the benefit of planning. And it would be a leveler of power, securing for world government greater independence from the most powerful nations. But of course it is just those most powerful nations whose assent would be a prerequisite for, and is inalterably opposed to, any such compulsory tax. Realistically, the prospects of funding environmental repair through a general world tax revenue are negligible.

The infeasibility of a general tax has led to calls for special sorts of global taxes designed not for the support of world government generally, but as problem-specific instruments for underwriting global environmental repair. An example is the Planet Protection Fund that the late Indian Prime Minister Rajiv Gandhi put forward in 1989. Under that proposal, each nation would contribute 1/1000 of its GNP, the proceeds ($18 billion a year, by his calculations) to be used to help Third World countries to develop environment-friendly technologies that would be given to them free of charge.[7] Gro Harlem Bruntdland of Norway vented a similar, GNP-based proposal at about the same time.[8] However, neither idea, nor more recent comparable proposals such as the United Nations' goal that each country give 0.7 percent for foreign aid, while understandably embraced by the poor, has won the support of the rich on whom it depends.

Special Taxes on Environment-degrading Activities

The fund-raising variant that is heard more frequently is to base a tax not upon each nation's wealth as such, but upon certain of its environment-degrading activities, most commonly the use of carbon.[*] In 1990 Brazilian president (then candidate) Fernando Collor proposed an international tax on carbon emissions at $100 a ton, with 2/3 of the $450 billion estimated revenues earmarked for developing countries, ostensibly to enable them to preserve and restore the environment.[9] At Rio the Italian environment minister proposed an "energy tax" of $1 a barrel of oil on all OECD members, the $25 billion to be used for Third World development.

Because carbon emissions, particularly those from transportation and industry, closely correlate with a nation's level of development, carbon taxes are to some extent a repackaging of the old proposals to tax national wealth. But not entirely. The carbon tax aims not only to produce a pool of wealth that would be available for transfer, but also to dampen the use of a major contributor to climate change. By raising the cost of fossil fuels to the users, the tax would be aimed at making the adoption of other, more environmentally benign alternatives, such as conservation or hydroelectric power, that much more attractive.

Now, we have already seen (chapter 6) that the carbon tax, in any of its several variations, falls far short of what is required to meet the second goal, as a deterrent to climate change. The benefits are entirely too meager. The United States currently emits 3.3 billion metric tons of CO_2 annually. The Environmental Protection Agency has estimated that if the United States were to impose a surtax on gasoline of $1.24 a gallon (doubling the pump price), we could expect an annual reduction of only 78 million tons.[10] An earlier study had sug-

---

[*] For purposes of this discussion, we need not distinguish whether the carbon-based funding mechanism take the form of a traditional tax or of a system of tradable emissions allotments, with funds coming from the auction of rights by the government.

gested that if the United States and indeed the entire OECD began now to tax all carbon heavily enough to double prices, by the year 2050 the "payoff" in lowered global temperature would be only about 1/10 of a degree centigrade.[11]

On the other hand, the tax looks better if we evaluate it less as an incentive to substitute energy sources (although as a constant reminder that we should be weaning from fossil fuels), and more as a mechanism to raise funds. Indeed, if, as I suggest, we get away from the preoccupation with climate change and recognize a slew of equal priorities, the meager effect on fuel substitution is not a critical objection. The question is, how does a carbon tax rate as a fund raiser help the world cope with loss of biodiversity, and so forth?

The principal objection has been voiced by the United States, along with the Saudis and some other OPEC (Organization of Petroleum Exporting Countries) members. It hardly seems fair, and may inject inappropriate incentives, to put the entire onus on oil. Why not include a tax on other GHGs, at the least coal if not methane?

THE GLOBAL COMMONS TRUST FUND

My own proposal takes this last line of thought a step further. I would establish a Global Commons Trust Fund (GCTF). The idea is like that of the carbon (or all-GHG) tax, but more broad-based on the revenue base side and more restricted on the spending side. Essentially, on the funding side, the idea is to capitalize on revenues from all commons-connected activities, and not only from charges for carbon "storage" in the atmosphere, the most familiar fund-raising scheme; on the expenditure side, the funds so raised would be applied to the conservation and repair of the commons areas rather than to distribute them back to individual nations to let them expend them on developmental projects of their choice, however tenuously connected the projects are to the environment.

Let me expand. The reason we criticize GHG emissions is not only the anticipated damage, but the fact that individual nations are saving themselves money—the costs of pollution abatement—by appropriating, free of charge, a valuable feature of the commons areas: the transport and storage capacity of the atmosphere. But of course there is the same seizure of a common-property right when nations dump waste or overfish the seas. Assuming the commons areas—the atmosphere, oceans, space, and so on—to be the Common Heritage of Humankind, the common property of all nations, one may argue that the users of the commons areas ought to be charged for their use. The charges would curtail the level of abuse and at the same time underwrite the expenses of repairing the damage we have already done.

The revenues, while difficult to estimate, are potentially enormous. Consider some rough projections.

The oceans.   Start with the oceans. The world harvests 175 billion pounds of marine fish annually. A tax of only one-half of one percent on the commercial value would raise approximately $250 million for the proposed fund. The same token rate on offshore oil and gas would add perhaps $375 million.

There is another, dirtier use to which the world community puts the oceans: as a dump site for waste. The official figures, almost certainly underreported, amount to 212 million metric tons of sewage sludge, industrial wastes, and dredged materials yearly.[12] A tax of only $1 a ton would raise $200 million more.[13]

The atmosphere.   Nations use the atmosphere as they use the oceans—as a cost-free sewer for pollutants. By burning fossil fuels and living forests, humankind thrusts 22 billion metric tons of carbon dioxide into the atmosphere annually.* A CO_2 tax of only ten cents a ton would raise $2.2 billion each year, thirty times the current budget of UNEP. Taxing other GHGs such as nitrous oxides at a comparably modest (dime-a-ton) rate, indexed to their "blocking" equivalent to CO_2, would bring the total to $3.3 billion.[14] The same

---

* The equivalent of approximately 8 billion tons of *carbon*, which can alternatively be the basis for a "carbon tax."

ten-cents-a-ton tax could be levied on other (non-GHG) transfrontier pollutants; a sulfur dioxide levy, for example, would produce $16 million.[15]

Space. Commercially tapping the riches of the planets may remain futuristic, but the rights to "park" satellites in the choice slots represent a potential source of enormous wealth right now. Most valued are points along the "geostationary orbit," the volume of space 22,300 miles directly above the earth's equator in which a satellite can remain in a relatively fixed point, relative to the surface below. The number of available points is restricted by minimal distances required between satellites to avoid interference.[16] Rights to spots directly above the earth's equatorial belt are also valued because they are exposed to exceptionally long hours of sunlight and are therefore ideally situated for production of energy from solar radiation, both as a support for special operations such as high-tech, gravity-free manufacturing, and perhaps ultimately for commercial redirecting to earth.[17]

The geostationary orbit being directly above their heads, the equatorial nations, in their Bogota Declaration of 1986 declared the orbital space to be among their natural resources, "an integral part of the territory over which the equatorial States exercise their national sovereignty."[18] While the world community has accepted the coastal states' extending their jurisdictions laterally outward into the sea, it has totally ignored the analogous and inconvenient claim of the equatorial states. (As my international relations colleagues would quip, although the equatorial states' reasoning may be as strong, their armies are feebler.)

As a result, frequencies in space, "the most precious resource of the telecommunications age"—worth to users an estimated $1 trillion globally over the next decade[19]—are now parceled out by the World Administrative Radio Conference (WARC).* WARC's assignment

---

* WARC convenes periodically under the auspices of the International Telecommunications Union. For the legal and diplomatic background, see Milton L. Smith, "The Space Warc Concludes," *American Journal of International Law* 83 (1989): 596–99.

is, admittedly, tough, but the approach it has adopted is folly. The slots are simply apportioned by formulas too cryptic for any outsider to follow and handed out free. The tiny island nation of Tonga, after being awarded three to six orbital positions gratis, turned right around and put them up for sale, recently striking a deal with a satellite company for $2 million a year "rental." And it is reportedly seeking more such deals.[20]

Why should the rights to any of these slots and spectrum positions, the legacy and province of all humankind, and worth trillions of dollars to users, be doled out like free lottery tickets, while those who would mend the planet are severely hobbled by a lack of resources?*

Biodiversity.   I am a little more ambivalent about including biodiversity as part of the Common Heritage of Humankind in the sense of making it a tax base for the fund. While many of the seas' riches lie in commons areas beyond any nation's jurisdiction, when we talk about dividing up biological riches, we mean resources that lie within established national territories. Of course, those who want to tap the biotechnological and pharmaceutical potential are not proposing to appropriate physical matter from a nation's forests. The hope is to copy and exploit *genetic information.* But that makes small difference to the biologically rich nations such as Colombia and Brazil, who regard global demands to share the good luck of their biological wealth about the same way the Saudis would react to arguments that the world should co-own its oil on the gounds that it is the Common Heritage of Humankind, and then having so much of it is pure luck, anyway. The proposal may simply intrude too far into the host nation's sovereign space and prerogatives, which is why the

---

* The same questions are being raised, incidentally, in the U.S. regarding comparable practices of the Federal Communications Commission. The policy of allowing the FCC to give away the radio spectrum, long a subject of academic ridicule, is coming under fire of some politicians and "market management" conservatives who recognize the enormous potential to cut the deficit if anyone were willing to confront the entrenched interests in broadcasting and their allies that are nested in and around the FCC, and insist on a government auction.

Rio negotiators rejected labeling biodiversity part of the Common Heritage in favor of the limper "common concern."

On the other hand, perhaps a compromise could be worked out whereby the industrial world's pharmaceutical companies, which will presumably manage the exploitation of the potential, pay a modest royalty into the GCTF (or perhaps even into the GCTF earmarked for environmental development in the nations from which the genetic information came).* In either event, whether or not we emerge including biological diversity as part of the resource base for purposes of a GCTF tax, projects to protect biodiversity would qualify for support under the GCTF as furthering the protection and repair of the commons.

And even if we do not include biological diversity in the "tax" base, the total thus far is over $6 billion a year. And that is before adding the yield of a surcharge on uneliminated HCFCs and other ozone-depleting agents, on toxic incineration at sea, or on the liquid wastes that empty into the oceans from rivers. Consider also fees on the minerals that someday will be taken from the seabed and, perhaps, depending on the staying power of the conservation movement that is fighting the prospect, the Antarctic.

Legal Charges

Another way to bolster the fund would be to designate it the receptacle for legal charges assessed under existing and envisionable commons-protecting treaties, such as those regulating ocean pollution. Consider, for example, the treatment of oil spills on the high seas. No individual nation can claim to have been damaged, but ocean pollu-

---

* From a legal perspective, there are three broad alternatives: (1) No automatic world system of intellectual rights (copyright) protection for the information. This is the default position that would leave every pharmaceutical firm, etc., free to exploit what it can acquire free of payment to owners of the original "resource," although each nation or landowner will undoubtedly respond by trying to condition (exclusive?) entry for scientific and pharmaceutical exploration on a compensation agreement negotiated with firms that make the best offer. (2) International standards protection for owners of the original "resource," so that in all cases the exploiters

tion agreements could easily be amended to provide that "damages" (perhaps according to a schedule, if measurement of actual damages is too conjectural) be paid into the trust fund to be available for general research and repair. There are illustrations of this technique domestically. After the spill of the highly dangerous pesticide Kepone into the James River in the United States, Allied Chemical, which was responsible, established such a fund for the James. More recently, in the wake of the catastrophic incident at the Sandoz factory in Basel, Switzerland, in 1986, Sandoz established a $10 million fund to further the ecological recovery of the Rhine. Part of the fund is being used to establish an interactive data base on the Rhine ecosystem.[21]

Objections on the Revenue-Gathering End

Every country would not blithely submit to the levies. One can anticipate resistance among the developing nations. But they do not face the highest charges, and the fund therefore does not depend on them. For the developed countries, the $6 billion or so is a lot more realistic than most of the figures that have been bandied around.

Some countries will object to any tax on activities within their territories or, in the case of the coastal states, within their self-proclaimed Exclusive Economic Zones (EEZs). But the charges are not for what nations do within their sovereign "insides"; they are aimed at the effects of their activities on the "outside" world. Moreover, many noncoastal, "landlocked" nations—along with many scholars—continue to regard as semilegitimate, at best, the coastal states'

---

have to pay the owners for value derived, even absent any special negotiated agreement. And (3) some standardized copyright-like protection and some "royalty" payment earmarked for the fund, as well, perhaps, as for the nations where the information is found. The biodiversity convention is closest to (2), exhorting countries using the genetic resources to pursue government actions "with the aim of sharing in a fair and equitable way the . . . benefits . . . with the Contracting party providing" them on "mutually agreed terms." Art. 15, para. 7. But other language suggests that resources might pass through an ill-defined "financial mechanism" consistent with (3).

proclaimed EEZs of 200 miles and more from their coasts. Allowing the coastal states to supply exclusive *management* across these zones makes a certain amount of sense; some management of ocean resources is better than none (see chapter 4). But allowing the coastal states to snatch all the wealth without any *accounting* to the rest of the world, just because it happens to be closer to them, is less defensible. That is why I would include taxes on resources such as fish and oil taken anywhere from beyond the traditional territorial boundaries of 3 or 12 miles from the actual coastline, and not merely from the smaller region of high seas remaining beyond their EEZs.

Many people will object to the pollution-charge component of the proposal, calling it offensive to permit pollution-for-pay. The answer is that some pollution is inevitable, and it is more of an outrage that we let the polluters get away with it, as they presently do, free of charge.

It is true, however, that the charges I have suggested are arbitrary. I offer them only as talking points. But it is not easy to provide figures that are not arbitrary in some degree. Theoretically, there are two approaches that might minimize arbitrariness and produce a "rationally" defensible set of figures. The first is to estimate the damages done by each of the various activities (where applicable),[*] and to charge a tax accordingly. Roughly: if incinerating a ton of dioxin on the high seas caused $500 damage, that would be the charge that the burner would have to pay.[22] The problem with this approach has already been reviewed. Principally, we really have only the faintest idea of the damage that a ton of ocean incineration or atmospheric carbon causes. Indeed, much of the harm we imagine comes from pollution, such as impairment of ecosystems, is hard to quantify even in principle and controversial to account for because of the long, generation-skipping time periods involved. (In the case of sulfite aer-

---

[*] A nation's use of space slots does not damage the world the way its waste emissions do. Not charging for the use of the slots amounts to forgoing an opportunity to redistribute wealth, but the world is not worse off on net, as it is almost certainly from continued pollution externalities.

osols, which act as reflectors counteracting global warming, there might even be net benefits.)

The second approach is to forgo any attempt to tie the charges to damages and instead gear them to what each user is willing to pay. We have already seen in chapter 6 how this could be accomplished. To review with a quick illustration, every year, hundreds of thousands of tons of synthetic organic compounds, including chlorinated hydrocarbon pesticides, industrial chemicals such as PCBs and organometals such as tributyl tin (TBT) are released—much of it to work its way into the world's oceans.[23] Suppose that the aforementioned Authority determines the contemporary figure in some region to be 1,000 tons, which it deems unacceptable. It determines that the loadings could be phased down in the first stage of withdrawal to, say, 500 tons. It would auction off 500 one-ton-per-annum permits. Each bidder would be forced to consider its alternatives. These would include substitution of "natural" pesticides, recycling, providing landsite fills for the waste metals, and so forth. Each bidder would come back and bid a price for the permits that reflected what the permit for ocean discharge rights was worth to it. Nations that had relatively better alternatives would take them and drop out of the bidding early. Conversely, nations whose safe landfill sites, for example, were filled would stay in and bid the price up. Everyone would benefit, even if indirectly, since the money that was put into the fund would be available for such ocean-protecting tasks as monitoring and insuring the integrity of the outflows and perhaps of the dump sites. We would "squeeze down" on the volume of ocean pollution, and pay for a safer, cleaner ocean than we have now.*

While I find much to favor in this system theoretically, we have already acknowledged the problems. For a permit auction system to achieve theoretical efficiency, the Authority would have to know the

---

* There is an interesting question regarding the pricing strategy to adopt: if the deep sea is virtually the only site for dumping certain material, should the managers pursue a monopsonist strategy, charging what the market will bear and thereby extracting as much as they can for the fund? I think not. The higher the charge for legal disposal, the greater temptation of waste disposers to cheat—indeed, even, if U.S. experience is a guide, to invite the participation of organized crime.

efficient level of loadings. And in truth, this is akin to, although not quite identical with, the problems of estimating the damage. We are no more omniscient about the "right" aggregate level of loading than about the "right" price per unit dumped. This is a defect that made us highly skeptical about auction systems as a way of achieving efficient levels of reductions. But here the primary focus is on fund raising rather than on abatement. The question is simply whether, considering the whole range of commons-using activities, there is not considerable room for modest, auction-driven restriction in many activities, well before we hazard inefficiencies, and indeed, well before we hazard raising more revenue that is politically acceptable to entrust to world government control.

Indeed, if there is a real objection to the proposed GCTF, it is that the initial rates I have suggested are probably too paltry. Viewed as a strategy for reducing environmental damage, the levels advanced for discussion—ten cents a ton for carbon usage—are highly unlikely to confront the polluting nations with the full costs of the damage they are causing to the global environment and therefore will fall short of inducing the "right" amount of conservation and pollution control. Viewed from the reverse side, as a strategy for maximizing revenues for the environmental infrastructure, they fall short of extracting the full value of what users would pay if they were required to bid for restricted rights at an internationally conducted auction. For example, to a nation seeking a site for waste disposal, the value of depositing wastes in the ocean or atmosphere is a function of the alternative costs it would face if forced to remove and dispose of them domestically. In the case of some wastes now being deposited in the oceans, the costs of disposal on shore are hundreds of dollars a ton. I do not pretend to know the right charges. What we should insist upon, however, is an end to the present practice, which allows the dumping free of charge, with no "compensation" paid to the global community.

I have quite frankly juggled my proposed rates to produce figures that lie roughly at the boundaries of what I suspect is financially realistic. Others might set their sights on other targets. The basic thing is to start discussing it.

Objections on the Expenditures End

In terms of where the money would go, once gathered, the GCTF is distinguishable from its rivals. The proponents of the carbon tax schemes, like the proponents of taxes on national wealth, have typically left the details of payout strategically inexplicit.[24] In general, the Third World is assured that the proceeds would be distributed to them; the environmentalists are encouraged to believe that, once in those countries, the money would be earmarked to meeting the challenges of industrialization in a still hazy "environmentally benign" manner—although, at the same time, Third World spokepersons rejoin that to attach any strings is an affront.

Under the GCTF proposal, funds would be restricted to underwriting only internationally significant efforts, those that most connect to the Common Heritage of Humankind. Thus, the expenditures, all of which originate from a "use" *of* the commons, would return *to* the commons. Funds would be available to improve global monitoring and modeling; to prepare adaptive strategies, such as the development of fast-growing (and carbon-withdrawing) trees; to inaugurate and police improved methods of waste disposal; to inventory, gather, and store genetic material; to underwrite the transfer of environment-benign technology to developing nations; and to promote energy conservation and general antiwaste behavior. The funds could underwrite developing institutional readiness to respond to various sorts of crises with the global equivalent of "firefighters." For example, no single nation can afford to keep on full-time alert a staff trained and equipped to contain oil spills with oil-eating bacteria, etc. No single nation anticipates enough incidents to warrant the expense. But a force with global responsibilities, financed out of the fund, might well be justified.* Similarly, to deal with nuclear acci-

---

* To counteract damage that the Persian Gulf environment suffered in the course of the U.N.-Iraq hostilities, the International Maritime Organization found itself trying to pull together a "Gulf Oil Pollution Disaster Fund," but the efforts were clearly hampered by the ad hoc nature of the need, and there were consequent delays in response. See "Middle East: International Maritime Organization Launches Gulf Oil Pollution Disaster Fund," *International Environment Reporter (BNA)* 14 (1991): 127–28.

dents one might want to have on call at least a crew of administrators with plans and power to assemble an emergency team on short notice. It has been estimated that $150 million a year would underwrite an effective worldwide system to give early detection to viral diseases—so that the next AIDS-type epidemic does not overrun us by surprise.[25] All these ideas have, I believe, intuitive appeal, and some congenial forerunners in the literature.[26]

There are objections. One can grant the legitimacy of raising revenues from the commons but yet maintain that the funds be divided among the world's nations without environmental or any other strings attached. The position is hardly radical. In 1970 President Richard M. Nixon, in proposing an extension of the coastal states' administration over their adjacent seabeds (all the way to the edge of the continental slope)—a development representing an enormous expansion of United States territory—had expressed willingness that some percentage of the wealth realized in the enclosed areas be set aside for the benefit of developing countries as a sort of quid pro quo.[27] And indeed, the proposed (but unratified) United Nations Conference on the Law of the Sea (UNCLOS) seabed mining provisions endorse some equitable sharing of the eventually hoped-for minerals royalties, "taking into particular consideration the interests and needs of developing States and peoples who have not attained full independence."[28]

Some such redistribution of the commons' wealth back to nations, rather than reserving it for commons cleanup, can find considerable support. Just as a practical political matter, many if not most national leaders, given the choice, are likely to prefer bringing marginal resources within their own unfettered disposal, rather than to leave them in the hands of more remotely accountable (and potentially self-serving) international functionaries. And the position can claim a legal as well as moral legitimacy. If each use, particularly a consumptive use, of the commons requires payment, why is it any more just that the payment be made for the benefit of the commons, rather than pro rata, to each of the nations as legal "cotenants"? If each nation is entitled to compensation for its share of the commons wealth, then nations that want to "leave" their moneys in the commons account, to clean up the seas or whatever, are of course free to do so. But if developing countries with the urgent needs of Nepal and Bangladesh prefer withdrawal of their shares to meet domestic requirements, why should they not have that right?

My best response to this question is predominantly pragmatic. When we examine the feasibility of dividing and distributing the wealth in shares among nations, we run into a new ethical morass: Is the distribution to be made to nations on an equal basis (Liechtenstein's share to be equal to Bangladesh's?) or is there to be an adjustment for populations? Or for need? And if need, how is need to be measured? These problems strike me as so morally and institutionally intractable that dedicating the money for the provision of international public goods may be the only practical alternative.[29]

But there is another alternative to the GCTF. In the absence of a world tax, it has been noted that royalties and charges from ocean resources, Antarctic minerals, and outer space activities could be used to secure world government (and not merely the commons areas) from the vicissitudes of voluntary contributions and assessments.[30] Hence, even if we do hold onto the proceeds for globally public undertakings,[31] there remains the question: On what theory should the funds be reserved for the repair of the environment, rather than to address other world problems, from U.N.-authorized peacekeeping missions to medical assistance and famine relief?

These are appropriate and difficult questions that merit further discussion. I am not confident that there is a compelling theoretical defense for the GCTF, particularly since I myself place famine relief and local health so high on the world's priorities. But one has to consider the alternatives, case by case. When tens of billions of dollars were needed for peacekeeping in the Persian Gulf, they were quickly forthcoming, without any trust fund. Often, famines are not prolonged as much by a shortage of relief funds as by belligerents who block supplies that pile up at docks and airports. The question in these situations is to be overcome more by diplomacy than by dollars. Nonetheless, I would gladly have the GCTF tapped for famine and disaster relief when other sources were not forthcoming. Such a use of the fund might be seen as in keeping with the fund's basic notions, inasmuch as some of the world's worst calamities, such as famines and droughts, are brought on or exacerbated by adverse planetary conditions.[32] But where many more unambiguously local problems are concerned—the dirty water and the bad health care—it seems to me that all in all we might do better to augment and redirect the efforts of the many agencies that already have a hand in those problems: the development banks and agencies, the U.N. agencies such as the FAO, the WHO, the cadre of emergency relief organizations mentioned in the last chapter, and the recently inaugurated Global Environmental Facility (GEF), described later in this chapter.

Ultimately, arranging the supply both of local relief and of global public goods—global peace, global public health, global environment, and so on—is a practical problem, one that has to look to established practices and institutions and the feasibility of various funding alternatives. UNEP and the infrastructure of government and nongovernmental organizations that attend to the global environment are, as institutions, still young and chronically starved for funds. We should be asking: Are they getting less than they merit? And if so, what can we do about it? In the total mix of fund-raising techniques, what special contribution can we expect from new resources which a mechanism such as the GCTF could tap?

FINANCING OPTIONS IN THE LIGHT OF DOMESTIC POLICY ALTERNATIVES

Of course, to say that the GCTF money would come from "use" of the commons is more than a little misleading. What is really meant is: it would come from nations for the use of the commons. No free dolphin-free tuna-salad lunches. Would it be—could it be made?—politically acceptable?

While my own view is to enter into international discussions in a spirit of mutual cooperation, for purposes of clarifying the options it is best to shift from the perspective of critics sorting out a theoretical array of financing alternatives, and adopt the vantage point of a particular nation-state advancing its own self-interest. Let us suppose that the United States credits warnings about some environmental problem: it wants to maintain or repair the ozone shield, retard the thickening of the greenhouse blanket, reduce pressures on living ocean resources—whatever.

All sorts of inquiries would precede such a decision. What is the value of, for example, a certain level of biodiversity, and how real is the likelihood that we will fall below it if we take no remedial action? How much money does risk reduction warrant over various time horizons? And so on. There are judgments to be made as to the likely irreversibility of the peril, its projected onset and schedule of damages, and the applicable social rate of discount.

Suppose, however, that this work is behind us. We have both assessed and valued the various environmental risks in the light of alternative budget demands, and have (rightly or wrongly) come up with some budget constraint. We are prepared to spend, say, $1 billion to deal with the problem at issue.[33] We will assume, too, that we are motivated by our own national self-interest—leaving for later any adjustments that we may wish to introduce from altruistic moral concerns (chapter 10).

What are the options for spending the $1 billion?

The questions we face include:

1. Why—and in what circumstances—may it be appropriate for protective and remedial regimes to take the form of unilateral, bilateral, or modestly scaled multilateral regimes, rather than of fuller-scale international cooperation?

2. In those circumstances where the ideal response *is* large-scale international cooperation, under what further circumstances is it preferable to institute environmental *funds* of various sorts (variously funded, variously governed) at the heart of the cooperative effort?

Unilateral Internal Responses

The first question then is, how most efficiently to expend the money budgeted. And the beginning of an answer is that, before we commit to institutionalizing and working through a large-scale multilateral fund, there are a number of alternatives that have to be considered.

The first alternative is to expend the budget unilaterally on inter-

221

nal problems, without any international entanglement. Put aside clean air, clean water, forest and wildlife programs; we are still trying to figure out how to decontaminate our waste dumps, dispose of spent nuclear fuel, and destroy chemical weapons stockpiles. These are deadly serious undertakings, deserve very high priority, and are best undertaken virtually alone.

When our concern turns, as it should, to problems beyond our frontiers, such as ocean pollution and climate change, there are obvious advantages of forming cooperative arrangements with others. First, other nations are jointly responsible for the problem, so that the United States, by unilaterally restricting its pollution or overfishing, etc., cannot be assured that the problem will be mitigated. Indeed, other nations may simply take up the slack. Second, other nations are jointly affected by the problem, so that unilateral efforts by the United States to resolve it may be accepted by other nations as a beneficial "free ride" with no mutual obligations attached.

But as we saw in chapter 7, while such cooperation has advantages, it also has its costs. Depending upon those costs, there is nothing inherent even in such a world-spanning phenomenon as climate change to make us renounce a unilateral internal response. What one fears from climate change are various symptoms—sea surges, windstorms, droughts—which are inevitably local to someone. Rather than make large-scale expenditures to prevent climate change from occurring, a nation may find it more efficient to wait thirty or forty years and adapt in a manner specifically directed to the local manifestations if and when and where they appear: putting seawalls in one area, improving water distribution in others, as required.

This is similar with regard to U.S. concern over loss of marine resources. Of course cooperation is in some sense ideal, and should be pursued. But it should be pursued with the knowledge that there is an alternative or supplementary go-it-alone option, if—the ideal cooperative effort failing—it comes to that. The improvement of marine resource management in our own EEZ and the development of local aquaculture may, depending upon the "costs" of securing cooperation, be a better strategy than arranging global collaboration on the high seas.

Indeed, there may even be side benefits of internally focused programs. Viewed from a narrow perspective, the administration's proposed 1993 budget for the United States Global Change Research Program, well over $1 billion, is an important source of support for U.S. universities and research institutions.[34]

Cooperative Internal Responses

An obvious defect of go-it-alone strategies is that, at the least, any nation committing $1 would like to see its contribution (certainly the share associated with external benefits) matched by other nations in mutually beneficial ways.

A nation may of course take internal action in the hope or expectation that others will follow its lead. If the chances of voluntary imitation are slim, the prospects of cooperative, "matching" behavior can be improved by recourse to formal, multilateral accords. Many international agreements take such a form, each party undertaking to implement jointly favored internal measures. An example is the Convention on Long-Range Transboundary Air Pollution (LRTBAP), already discussed, under which each signatory commits to reduce SO_2 emissions 30 percent from 1980 levels.

"Matching" gestures can go beyond coordinated reductions in emissions. Nations can—and do—agree to harmonize their tax rates,[35] environmental laws and enforcement, pollution equipment, and many other things. And note that none of this requires any great global management structure or even flow of funds.

What advantage might a fund display over such tit-for-tat internal measures? One might suppose that funds lend themselves to the fostering of cooperation by introducing a measure of flexibility. By establishing and making payments through an international fund, each nation can ante-up small incremental contributions of the full amount that it is ultimately willing to put forward under conditions of full cooperation. That way it can withhold full donation until it has received more information, both about the state of the world and about the willingness of others to advance the amounts it considers prerequisite to its making a larger "play" on its next turn to donate.

At each move, it signals its willingness to engage in conditional cooperation if others keep their end of the deal.

This happens. For example, at the London ozone negotiations in 1990 the United Kingdom contributed $9 million to the ozone fund immediately, with the pledge to raise its contribution to $15 million if China and India were to sign on.[36] That maneuver was intended as an inducement to (and perhaps a source of third-party pressure upon) the most crucial nonsignatories as a way of increasing the likelihood of realizing the fully cooperative outcome—and it worked, both China and India ultimately consenting.

On the other hand, funds are not unique in providing opportunities to signal conditional cooperation. Each nation can offer as its "move" its domestic commitments or its progress in meeting emission reduction targets. Indeed, anyone carrying out a comparative analysis of signaling strategies would have to consider how a commitment to donate to a global fund compares with other sorts of signals in terms of strength and reliability. A national pledge to support a fund is presumably subject to annual review, and therefore insecure across domestic budgets. By contrast, some "internal" gestures, such as the dismantling and replacement of dirty generators, represent sunk commitments which might be regarded by other nations as stronger, more persuasive signals of cooperation than pledges to make annual contributions to a fund.[37]

Bilateral Arrangements with Resource Transfers

Such reciprocal mirroring of internal responses will not be satisfactory in certain situations. If Nation A is determined to preserve black rhinos or elephants, etc., and it doesn't have any black rhinos or elephants (or suitable habitats) within its own borders, then it has no choice but to make arrangements with a nation that does. In other words, the benefit sought may necessitate one nation paying another nation to do something.[38]

In these circumstances, it might be advantageous for Nation A to arrange payment through the intermediary of an international fund—benefits in terms, for example, of producing matching funds from other countries or benefiting from the shared overhead of technical expertise. But note that for A to get what it wants certainly does not *require* internationally funded cooperation. Nation A may do better to seek out, itself, nations with promising rhino habitats and, playing the one off against the other, arrange the best conservation deal, dollars-for-rhino, that its money can buy.

Such direct nation-to-nation conservation agreements are being made all the time.[39] Indeed, the amounts presently committed by the United States in direct bilateral conservation deals probably well exceeds the commitments through international channels. That would be in line with the reported policy of the Reagan administration, which was to concentrate its limited environmental protection funds to bilateral aid programs administered by the Agency for International Development.[40]

The role of such bilateral compensation is not limited to cases where Nation A utterly lacks, and Nation B possesses, some resource or service A desires. There are opportunities for gains from trade wherever the same "product" can more efficiently be provided by one nation than by another. Suppose that Nation A, after years of improving its antipollution technology, has reached a point where each additional $1 spent on pollution abatement will eliminate one unit of pollutant from an international air basin. In less developed Nation B, A's neighbor, the $1 will eliminate five units. In such circumstances, it may be advantageous for A to divert its pollution reduction budget to B, on condition that B install the best available technology in its factories. Indeed, we have already explained West Germany's contribution to the modernization of East Germany's noxious Buna Chemical complex on this very basis.[41]

Multilateral and Regional Arrangements

In regard to ocean protection, the preference of UNEP for a Regional Seas Program, with its area-by-area focus, has already been mentioned.[42] Even in regard to problems such as climate change, which are "global" in the sense that their impact is likely to affect every region of the world, it does not follow that responses need be

225

"global" in the sense that all 140-odd nations need join in. Seven countries—the United States, the former Soviet Union, Brazil, China, India, Germany, and Japan—account for well over half the additions of greenhouse gases; or, put otherwise, an agreement among those seven nations to reduce their emissions 25 percent would result in a 12 percent reduction worldwide.[43] And, again, all this could be done without the entanglements of a full-scale international agreement, if the benefits do not merit the effort. And it all could be done without the intervention of my, or anyone else's, version of a global fund.

The Virtues and Vices of Global Funds:
The Donors versus the "Democrats"

Although much can thus be done outside any permanent global funding mechanism, environmental diplomats around the world are increasingly calling for them to play a key role in environmental defense efforts. Most prominent is the Global Environmental Facility (GEF). The GEF, first proposed by France, was established in 1990 under a still ambiguous plan for joint management by the World Bank and UNEP. As of 1992 thirty-five countries had joined and it has gathered $1.3 billion—a small fraction of its announced requirements.

Hardly sooner than GEF had been born, and well before it had had a chance to acquire many distinctive features, the fund fell under an attack from the LDCs withering enough to discourage anyone but a certified romantic from advancing any more global-funding schemes. The charges include, first, that the GEF is elitist. Membership requires a minimum $4 million contribution, although the World Bank pledged that if a poor nation could only come up with two million dollars, the Bank would match the other half. "But can a less developed country whose people are dying of starvation contribute two million dollars to save the environment?" an Algerian diplomat responded.[44]

GEF is also called undemocratic, having been set up "privately, secretly," in a "hasty manner"[45] and, more irksome, with a voting structure weighted to give power to those who supply the funds, not those who will receive them. (This is the very point of working through the World Bank from the developed nations' point of view.) GEF's inferred mandate (it hardly can be said to have a jelled mandate) is criticized as being too attuned to global problems such as climate change, with inadequate emphasis on the local and regional plights more pressing to the Third World.* The World Bank is said to be a tool of the West and, in any event, cannot be trusted to manage the environment because its lending projects bear blame for much of the problem in the first place.[46]

For these reasons, among others, the ministers from fifty-five developing countries, meeting in Kuala Lumpur, endorsed the idea of a Green Fund presented by the Group of 77 and China in the preparatory work for the UNCED meeting at Rio. The proposed counterfund would "provide for funding of activities according to the priorities and needs of developing countries"; it would "be democratic in nature, with an equal voice for all parties"; and it would "provide access and disbursement to all countries without conditionality."[47]

At this point, there is a standoff. Many LDCs want a no-strings-attached Green Fund. The developed countries are hesitant about funneling much into the GEF as long as the governance structure remains unclear, and will not give a lot (relative to Agenda 21) even if their voice is secured. Certainly they will not endorse a Green Fund. Is there enough potential benefit in a fund that it is worth the effort to try to bring the sides together? We have already touched on some of the advantages and disadvantages of international cooperation in general. Let us examine some of the general vices and virtues of global funds, in particular.

* Amid all the sound and fury over the need for more money, few seem to have noticed that there appears to be no backlog of meritorious specific proposals for spending. As of 1991, the EPA official who represents the United States on the ozone fund's executive committee, Eileen Claussen, was said to be "dismayed by the quality of $7.5 million in spending proposals developed by the United Nations Environment Program and the United Nations Development Program." Larry B. Stammer, "Ozone Aid for Third World Slow to Arrive," *Los Angeles Times*, April 15, 1991, 1.

First, the Vices

Put aside the costs of overcoming negotiating conflicts, which trouble any multilateral negotiation and are not specific to funds. From a negotiant's perspective, one thing that distinguishes multinational funds in particular are the administrative costs that, while shared with others, are almost certainly harder to contain than what one's own bureaucrats will squander. The United Nations and other international governmental organizations have not gained a consistent reputation for tight-fisted management practices.[48] There is no reason to suppose that a fund, GEF or Green or GCTF, will do much better. In establishing the fund called for by the Montreal ozone protocol, the Montreal Protocol Interim Multilateral Fund (MPIMF), UNEP was left with a mostly symbolic and supportive role because as administrator it would have taken 13 percent off the top as "administrative support charge"[49]—hardly the only item of administrative overhead the ozone fund is going to face. These cost considerations partially explain the fact that the United States is willing to budget over $2.5 billion for its own environmental efforts, and why Japan is establishing its own (Japanese-managed) $1.6 billion fund for global environmental projects.[50]

But the United States and Japan and the other potential "big spenders" have another reason to be hesitant, independent of management expenses. From any contributing nation's perspective, there is loss of control over agenda. The larger the control group and the less the fund is restricted to specific undertakings well defined in advance—such as storage of agricultural germplasm, the preservation of named world heritage artifacts or wildlife, etc.—the less assurance there is that the agency's judgments will conform to any member's own preferences. In today's geopolitical climate, it is not unreasonable for the major prospective contributors to fear that environmental priorities will be subordinated to other goals, specifically to foster Third World economic growth, even if the link between industrialization and environmental degradation cannot be broken.[51] Indeed, even if the fund managers make what the contributor would consider the "right" disbursement, the control loss problem does not end, in that the power to control any misapplication of the funds in the hands of beneficiaries is one more tangle in the agency's politics. (Assuming here, contra the Kuala Lumpur Declaration, that the concept of "misapplication" has meaning.)

At the extreme, the fund's agenda may be captured by groups who want to transform the institution's focus into political issues that are utterly unrelated to the members' original intentions. The nation that finds itself "entrapped" in an inhospitable institutional structure has the option of withdrawing participation and support (as the United States did with UNESCO). But while it is always possible to withdraw from the institution, it may not prove so simple removing oneself from the institutional power one has helped to create. There is an arresting illustration of this in the context of the ozone negotiations. The Ozone Protocol signed at Montreal in 1987 wisely prohibited any signatory from trade in listed agents (CFCs, etc.) with nonsignatories. An unforeseen result was that in 1990, when the signatories reconvened in London to tighten up the protocol, and, in particular, to accelerate the phasing out of CFCs, the U.S. chemical companies, while wary of many proposals, realized that if the United States were now to defect, U.S. firms would face an embargo of sales to signatory nations—hence, a loss of billions of dollars in international trade. The companies themselves thus became a force in pressing the U.S. government to work within the framework.[52] Once committed, the United States had found the costs of extricating from the convention too high.

The Strong Points of Multinational Funds

Despite these weaknesses, international funds have a significant and growing role to play in the world community. Any fund, to be successful, should be designed to occupy areas where the strengths dominate, and to do so in a way that works around the flaws.

THE LEGITIMATING FUNCTION

In some circumstances, the sponsorship of an internationally funded organization, even if not required financially, is required to legitimate the funded activity. We have already advanced as an illustration several commons-affecting cases, such as the proposal to nourish algae colonies on the high seas with iron ferrules.[53] But an internationally funded body can provide comparable services even where the commons areas are not involved, but the task involves integrating activities internal to a number of nations. For example, the establishment of an early warning system to detect emerging viruses such as AIDS might call for establishing monitoring systems in many nations, particularly those including or bordering rain forests.[54] The task of collecting and collating the data would presumably be advanced if an international organization (such as WHO) were in charge.

Indeed, wherever there is an international environmental agreement in place, monitoring for noncompliance may have to rest upon some international body. For example, the Montreal Protocol includes a clause restricting "nonessential uses," presumably aimed at preventing an LDC from using CFCs for aerosols or as a foaming agent, even during the period in which full application of the convention to LDCs is deferred. It has been suggested that embassies of industrialized nations certify compliance,[55] but an international organization, governmental or nongovernmental, might perform the service more acceptably.

SOURCE-DRIVEN CONSIDERATIONS

An international fund may be inherent in the source of the funds. If we implemented a plan to raise revenues by charging nations for their use of the global commons (e.g., a tax on ocean minerals or fish), as a practical matter an international fund would be the most feasible beneficiary.

COUNTERACTING PROBLEMS OF STRATEGIC BEHAVIOR

I have acknowledged that a fund is not the only, and not always the best, response to the full range of strategic behavior that may be expected in the amelioration of global environmental threats. Nonetheless, there may be a subset of problems for which a fund plays a key role. The provision of what Michael Taylor and Hugh Ward call, in a provocative article, *lumpy public goods* may be one of them.[56] Those authors point out that the supply schedule for many public goods, for example fire and police protection, is linearly related to the resources applied; in other words, the protection we get increases rather smoothly with the budget we provide. But the supply schedule for other public goods is "lumpy": if a bridge to span a river costs $100 million, we do not get 1/10 the value of the bridge for $10 million. No benefits are realized until the full costs are expended.

Projects resembling that bridge budget emerge in the context of the global environment, where responses to problems turn out to be both very costly and lumpy: sometimes no amount of the public good can be provided until some large threshold of expenditures has been passed.* For example, an atmospheric problem may become so severe that (as has been suggested) remediation may require construction of a huge space station, equipped with laser guns to destroy CFC molecules.

Taylor and Ward suggest a set of assumptions that would produce from the circumstances described the game of "Assurance." In Assurance, as contrasted with Chicken,** all nations would benefit from the project, with costs being shared; but either no nation has adequate resources, or none feels a high enough level of indivisible threat, to be bluffed into underwriting the full project independently. In that case, cooperative (shared) funding is not only the ideal solution; it is hard to envision any solution that is not based upon multilaterally funded cooperation.

Theoretically comparable, if less dramatic circumstances may explain the preference for some of the existing multilaterally funded projects. That is, there must be situations in which the multilaterally

---

* Note that this is distinct from the highly contrived asteroid threat which required the cooperation of *every* country and thus could be mapped onto the game of Weakest Link, chapter 7.

** If the situation is very severe and the requirement for remedial action highly urgent, the dominant strategy for each of several nations may be to fund the project independently; then the dilemma takes the form of the well-examined game of Chicken exemplified by a nonlinear warming cataclysm, such as the ice-pack melt discussed in chapter 7. In that case, the cooperative, mutually advantageous outcome may be thwarted by the possibility that each nation, miscalculating that, because it is rational for every nation to "go it alone" if need be, some other nation will underwrite the effort. And everybody, in our example, winds up facing inundation.

funded agency cannot only serve as a legitimating project manager, but as a sort of escrow agent in which nations can be encouraged to transfer incremental, matched contributions to the project, thereby aggregating the funds no nation individually could or would be inclined to provide.

LONG-TERM PLANNING

There are related situations in which the critical factor is not the magnitude of the resources per se, but the steady availability of resources over a long period. In other words, tit-for-tat, short-term funding may work for short-term projects such as, conceivably, cleaning up an oil spill. But where the nature of a project requires steadiness of administration and stability of funds over a long-term period, for example, a massive R&D venture, an international fund may be superior.[57]

OBJECTIFYING CONTRIBUTIONS

As indicated above, it is true that nations can cooperate by mutually beneficial, tit-for-tat sacrifices in kind: my stronger-air-quality law for yours. But because contributions of this sort are in kind, rather than in money, evaluation is unclear. Dispute over the value of each party's respective "contribution" may lower the aggregate benefits beneath the cooperative optimal.[58] A money fund, being somewhat more objective, may reduce these conflicts.

FORTIFYING OBLIGATIONS

More thought needs to be given to the relative enforcement advantages of agreements in which an international organ, as compared with a single nation (as in the case of a bilateral agreement), is the promisee. Consider first enforcement difficulties in bilateral agreements. If Nation B's promise to maintain a reserve is made to the United States directly, then B may later be inclined to renounce its obligations in response to "U.S. aggression" somewhere in the world. By contrast, if B's promise runs to an international organization, it might prove less fragile, since B's breach would be a wrong against the international community. Granted, this is not necessarily so—

which is why more thought needs be given. Many nations reliant, say, on the United States for trade assistance might be less inclined to breach against the United States than against an international fund.

INSURANCE AGAINST LOSS FROM BREACH

Related to the above is another reason why both payers and payees might favor an intervening fund over a series of bilateral agreements of comparable magnitude. Imagine a series of cooperatively mirrored transfer agreements running (bilaterally) from A to B, C to D, and E to F. If A breaches, B is totally at a loss. If, however, A, C, and E promised their funding (in effect) to B, D, and F, the fund, through arrangements among the nondefectors (payers C and E, and payees B, D, and F) is in a position to spread, and cushion, the shortfall. No one faces the prospect of being entirely stranded from A's defection.

CONCLUSION

Global cooperation does not require global funding efforts, either from the perspective of the world's health, or the narrower perspective of individual nations who enter negotiations with the most "self-interested" stance. The issue will become more acute for environmentalists, the more the LDCs demand control over the funds and insist that available resources be made available for development—no longer even "sustainable development"—with "no strong attached."

With the developed nations concentrating increasingly on their own internal problems and, when they look beyond, on the resuscitation of the Eastern bloc, the Global Commons Trust Fund is, I think, a viable alternative that the environmental movement would do well to rally around. It is not merely a roundabout scheme to take wealth from the rich nations and redistribute it to the poor. It simply seeks from users of the global commons a reasonable fee for their use of the commons areas, the proceeds to be earmarked for commons maintenance and repair. From the point of view of the rich nations, whose support would be critical for the GCTF's implementation, this solution should be generally more acceptable than many alternatives that are being proposed. From the perspective of the poor countries, a GCTF has the advantage of tapping funds not as a matter of largesse, but as a matter of obligation—something that the LDCs have long considered important. The users of the commons would pay for their use because they have a duty to do so, a duty that cannot in good conscience be turned on and off.

The GCTF in general meets the criteria for the circumstances in which a truly global fund is indicated. Many protective and remedial activities on the commons—such as restrictions on fishing, monitoring, and so on—demand the legitimation of a world rather than a bi-or multilateral mechanism; the funds being drawn from the commons, a trust for the use of the commons is the most intuitively appealing as well as practical alternative; and commons-connected problems present opportunities for strategic behavior for which an ongoing fund may be the best antidote.

By focusing on repair of the commons rather than on general development, the GCTF would be leaving many serious needs to the IMF, the World Bank, and the established panoply of international agencies, just as we leave world peace to the United Nations and its Security Council. But it would bring in new, and not necessarily "competing" funds for purposes that all humankind can agree are vital.*

---

* The fact is, there is no telling how much the new charges nations would face for use of the commons would reduce their contributions to other global—for example, United Nations—activities.

CHAPTER X

The Spiritual and Moral Dimensions
of the Environmental Crisis:
Of Humankind and Gnats

THE SPIRITUAL ROOTS OF THE ENVIRONMENTAL CRISIS

WHAT HAS brought us to this pass? Most of the book has been spent examining issues of institutional reform, on the theory that however we got here, the question now is what can we do about it. Some will say that this gets our priorities reversed, that to know how to cure something, we first have to identify and root out the cause. But like so much other wisdom, this is not as true as it sounds. We successfully treat lots of conditions without understanding, at any especially deep level, what caused them. Doctors could spend considerable time seminaring over a patient, his life imperiled by clogged arteries, whether the cause was lack of exercise, diet, genetic predisposition, or tension on the job. But so far as the patient's life is concerned, the exact cause may be quite irrelevant to the treatment. If he needs a bypass, he needs a bypass, and probably wishes they would stop talking and just get on with it.

I feel that way about many of the "deep," soul-scrutinizing discussions of our environmental problems. I don't know how to explain why so many canisters of wastes and weapons lie corroding on ocean floors. Fear? Distrust? Aggressiveness? Surplus capital? One only wishes we would recognize the threat and take care of it.

It is true that large-scale efforts to eliminate many problems benefit from an expanded appreciation of cause. But it is not always so. It depends whether the cause is anything we can do something about. Suppose someone could produce good evidence that the "cause" of the environmental condition was (in some historically satisfactory sense) *human greed* or some other such fundamental human spiritual flaw. What would it tell us about the cure? Well-entrenched human values and patterns of behavior resist reform. Consider how difficult it is to get people to change their sexual behavior in the face of AIDS, or to cut out cigarettes in the face of cancer and heart risks. When it comes to repairing the environment, we are faced with motivating people to alter patterns of behavior in the interests of risks that are not only uncertain but are far less *personal*: some of the severest perils are displaced onto future generations and the nonhuman environment.

There is another reason why I save the spiritual and moral crises for last. Individual human will does not have the force it once had. At least since the beginning of this century, the other startling population explosion (other than the fecundity of humankind) has been the proliferation of bureaucratic institutions. Increasingly, the actors that shape the modern world—that pollute it, clean it up, or do most anything else—are manufacturing corporations, banks, agencies, legislative bodies, pension funds, and so on. These institutions are of course comprised of individuals. But the genius of bureaucracy is to structure its individuals in a manner calculated to subordinate the individual will and spirit to the institutionally defined goals. A modern manufacturing plant turns people with personal concern for their children and the environment into workers who will cast out chemical wastes and weapons. Hence, even if we were to effect some changes in individual values, we would still be left with the task of transforming institutions.

Thus, my first instinct—reinforced no doubt by being a lawyer— is to place principal emphasis on reform of law and social institutions, and steer clear of human spirit. Hard as social institutions are to change, they are a lot more malleable than human nature.

But that is not a satisfactory response, either. Even if institutions are not transparent to the individuals who labor in them, and even if the present landscape appears so predominantly *organizational*, we cannot escape the ultimate responsibility of humankind in creating those institutions and calling them to account.

Thus, it is worthwhile to examine some of the conventional cultural and religious explanations before proceeding to examine what I consider the ultimate ethical and "spiritual" dimensions of our predicaments.

Blaming Culture and Religion

It is common to point out that different cultures, perhaps particularly in their religious dimensions, espouse radically different attitudes toward the environment. And from this it has proved a short, popular and attractive step, particularly in the United States, to blame Judeo-Christianity for having displaced an earlier, typically more pantheistic paganism.

The argument begins typically with a comparison. Taoism and various of the Amerind religions are held out as worldviews that idealize nature and even resist placing human beings at an especially revered center. By contrast, the Judeo-Christian tradition is launched in Genesis with God's determination to make man in His image, endowed with "dominion over the fish of the sea, and . . . the fowl of the air,"—the rest of it to be, in short, "dainty dishes" laid at our table.*

For the ancient Greeks, as the late U.C.L.A. historian Lynn White wrote, the world was quite different than what Christianity would bring:

> Every tree, every spring, every stream had its own *genus loci*, its guardian spirit. . . . Before one cut a tree, mined a mountain, or dammed a brook, it was important to placate the spirit in charge of the situation.[1]

Arnold Joseph Toynbee had noted the same contrast:

> In popular pre-Christian Greek religion, divinity was inherent in all natural phenomena. . . . Divinity was present in springs and rivers and the sea; in trees . . . in corn and vines; in mountains; in earthquakes and lightning and thunder. The godhead was diffuse throughout the

---

* See the selection from the Talmud on the dedication page.

237

phenomena. It was plural, not singular; a pantheon, not a unique, almighty super-human person. When the Greco-Roman world was converted to Christianity, the divinity was drained out of nature and was concentrated in one unique transcendent God.[2]

Another writer, a Chinese landscaper, points out that

> In the Christian tradition . . . holiness was invested not in landscapes, but in man-made altars, shrines, churches and basilicas that dominated the landscapes. . . . In the Christian view it was not emanation from the earth but ritual that consecrated the site; man, not nature bore the image of God and man's work, the hallowed edifice, symbolized the Christian cosmos.[3]

It is difficult not to find all this fascinating. But it is not especially convincing as a historical explanation of ecocide. To begin with, the biblical attitude toward the nonhuman environment is at least ambiguous. Vice-President Al Gore is the most recent public spokesman for the interpretation that "'dominion' does not mean that the earth belongs to humankind; on the contrary, whatever is done to the earth must be done with an awareness that it belongs to God."[4] Such an emphasis on a kinder, gentler Bible traces back, at least, to Philo.* Besides, even on the least generous interpretation of scripture, the charge is entirely too optimistic about the powers of the Bible to influence conduct; the Bible has not had the power to make us love our neighbors or turn the other cheek. Why then suppose it so controls our attitudes toward the environment?

Moreover, the biblical explanation ignores—or worse, overromanticizes—the environmental record of primitive cultures whose attitudes were "untainted" by Genesis. The reality is at best mixed. The Native Americans, before they got horses and rifles from Europeans, set immense, landscape-altering fires to enhance food gathering and

---

* Philo, celebrating the biblical injunction not to cut down an enemy's trees during siege (Deut. 20:1920), comments: "'For why,' [the law] says, 'do you bear a grudge against things which though lifeless [sic] are kindly in nature and produce kindly fruits." *Philo*, trans. F. H. Colson (Cambridge, Mass.: Harvard University Press, 1968), vol. 8, 149.

harvest buffalo. The aboriginal builders of the famed Easter Island statues deforested it of palms, destroyed habitats. The Anasazi, one of the most advanced pre-Columbian civilizations, made their environment their worst, and ultimately conquering, enemy.[5]

Long before Christianity, Professor White's vaunted "pagan animism" had proved itself a feeble protector of the Greek countryside. In the *Critias* Plato sadly reminisces on the disappearance of a green Greece that Man had long since made, even by Socrates' time, a memory: "What now remains compared with what then existed is like the skeleton of a rich man, all the fat and soft earth have wasted away, and only the bare framework of the land being left." In the same vein, in China, Tao and *feng-shui* notwithstanding, deforestation has been notoriously ruthless.* Part of the explanation is, as I have said, simply human needs—to eat and cook and keep warm. But we have to account, too, for the fact that a culture is a blend of many elements, often conflicting. Taoism has had to compete with other ideals that favored, as a "male" principle in (human) nature, the emulation of magnificent deeds, such as imperial-scale earth-moving projects that transfigured the landscape.[6]

In summary, cultural sentiments may induce the tree cutter to utter an expiatory prayer but are not as likely to stop him from cutting the tree. My impression is once more in sympathy with René Dubos, that "all over the globe and at all times in the past, men have pillaged nature. . . . If men are more destructive now than they were in the past, it is because there are more of them and because they

---

* *Feng-shui*, literally "winds and water," constituted China's most influential pseudoscience, a system of divination for determining the auspicious siting of human dwellings. "The general effect . . . [was] to encourage a preference for natural curves—for winding paths and for structures that seem to fit into the landscape rather than to dominate it; and at the same time it promoted a distrust for straight lines and geometrical layouts." Y. F. Tuan, "Discrepancies between Environmental Attitude and Behavior: Examples from Europe and China, *Canadian Geographer* 12 (1968): 176–91. Y. F. Tuan suggests that *feng-shui* may have contributed to "the short life of China's first railway," but agrees that its philosophic-religious line on the environment did nothing to prevent deforestation by needy populations. See also E. F. Murphy, "Has Nature Any Right to Life?" *Hastings Law Journal* 22 (1971): 467–84.

have at their command more powerful means of destruction, not because they have been influenced by the Bible."[7]

Yet, while attempts to portray Western religion as the villain fail to persuade, the thesis can be stated in a more moderate form. Toynbee is closer to the mark when he goes on to write that Christianity (and, I would add, Renaissance humanism) "made it possible to exploit nature in a mood of indifference to the feelings of" other creatures and natural objects. "Monotheism, as enunciated in the Book of Genesis . . . removed the age-old restraint that was once placed on man's greed by his awe."[8]

In other words, even if it is extreme to blame Judeo-Christianity for the condition of our environment, it is not unlikely that it has made cultures that embrace it less sensitive than they would be otherwise to the dark side of what they are doing. Buddhism, for example, *does* offer a more balanced mindset regarding humankind's relations with its environment.[9] And this Western *insensibility*, whether rooted in Christianity or anything else and wherever found in the world today, East or West, does call out for attention that can only be called *spiritual*.

Capitalist Greed

Judeo-Christianity is only one of a number of popular ideological villains. "Capitalism" and "capitalist greed" have been perennial favorites. The appalling environmental record of the former Soviet bloc has now become so notorious that the most strident and silly charges will probably fade. Indeed, so far as protecting the environment is concerned, one would suppose that the organization of market economies around price signals makes it considerably easier for a Western capitalist economy to sensitize its firms to pollution externalities than it is for a communist government to control communist firms. In market economies, if the political will is there, all the authorities have to do to tame polluters is squeeze down on their profits by escalating the level of punitive damages and fines.[10] Society's distaste for pollution can be readily translated into monetary terms—the firm's mother tongue. Eastern bloc experience has taught that communism, relying largely on more cumbersome bureaucratic control, lacked both the will and the technique to bring some of the world's worst polluters to heel.

And, ironically, the efforts in the Soviet Union may have been blunted by a self-deceptive ideology. Consider one item from the "capitalist greed" literature that is certainly worth preserving: the Soviet commentator I. Frolov's explication of the party line that only under communism could nuclear energy be developed safely, "since under capitalism, as a result of a striving for profits and toward an imperialist domination, the safety measures that are needed at nuclear power stations and other energy-producing facilities are often disregarded. . . . In short, [the reluctance to expand nuclear generating stations] is essentially not a matter of the peaceful use of atomic energy itself but rather of the dangerous forms such use acquires under capitalism."[11] The punch line is that Frolov's prescription for nuclear safety—generating stations will be safe if we can just place them in the hands of communist bureaucrats—was published only six years before the accident at Chernobyl.

And yet, as with the easily exaggerated and misfocused claims about religion, the censure of "capitalist greed" can be reformulated to bring out a truth in the underlying intuition. The right target is less capitalism than consumption. The life-style to which people are aspiring in all advanced and advancing societies, whatever the mix of markets and politics in their economies, is a significant part of the problem. In LDCs, people bring their own reusable bags and even refillable bottles to markets. As so many environmentalists claim, progress has become synonymous with supermarket packaging, big cars, pesticide harvests, take-out meals with their throw-away plates, hair sprays, and mountains of nonbiodegradable trash.

Few critics want to turn the clock back and model all human life on a rural Mexican village. On the other hand, some "primitive" practices are quite tolerable, and accepting them or not is simply a matter of custom. In many parts of England and Germany—their "development" notwithstanding—it is common to bring one's own tote-bag for shopping. Moreover, there are middle ways, and tastes can be changed. With increased sensitivity about energy and the environ-

241

ment, people are coming to regard large gas-guzzling cars and tiger-skin rugs less as coveted status symbols than as vulgar emblems of insensitivity and decadence. In the space of a few decades, the sight of a caged parrot for sale in a pet shop has become less droll than monstrous. There is increasing evidence that, across the world, buying habits in many markets are adjusting to reflect environmental awareness, even if it involves paying a premium for, "dolphin-free" tuna and biodegradable containers. The mechanisms that bring about these changes are subtle. Education, one that tutors in both fact and spirit, is as vital in the process as law.[12]

THE MORAL DIMENSIONS

There is another gap in coverage that must now be filled. I have been stressing biospheral degradation as it appears from the viewpoint, largely, of the law, with some recognition now of a cultural and spiritual backdrop. But a comprehensive perspective on control efforts requires a synoptic vision that adds the viewpoint of morals. Indeed, law, where it is to be effective, has always to draw on morals. But this interdependency of law and morals is especially crucial in the area of international cooperation.

In part, the special burden that morality must carry in the international field owes much to the absence of a strong central world government with powers, ultimately, of coercion. Treaties can and do raise the specter of sanctions. But in the near forseeable future we cannot expect even the most muscular treaty-made law to be backed by the familiar threats that domestic law deploys against polluters, such as criminal fines, punitive damages, much less imprisonment of serious wrongdoers. In fact, as we saw, it is likely that the more effective and threatening the drafters of a proposed convention make its legal sanctions, the dimmer will be its prospects of widespread ratification.

All this makes cooperation in the international arena all the more dependent on a feeling of rightness than on force. There is no world body with power to force a nation to protect its wetlands. To secure its cooperation, all the outside world has at its disposal is the threat of informal sanction—of labeling the noncooperator a bad world citizen; that and, one would hope, a shared sense that the world community's action in demanding wetlands conservation was morally justified—or, at least, *not unjust.*

But criticizing national behavior morally raises some tough questions. To begin with, there is a whole school of thought in international relations that denies that moral terms such as "just" and "right" have any application in this context. Realism takes many forms, from Hume and Hobbes to Hans Morgenthau, but the common elements are essentially (1) that global relations are international relations, predicated on the nation-centered system, and (2) that when nations conflict in the global arena, their actions are beyond good and evil, or (what amounts to the same thing) that any "right" that can be spoken of comes directly down to "might" and self-interest.[13]

My own view is that realism exaggerates, somewhat—and distorts the real issue. No one is so naive as to suggest that there is a substantial catalog of moral imperatives so overwhelming that each nation should, much less will, subordinate national self-interest to it in all circumstances. (Few moralists claim many overwhelming and unexceptional edicts in ordinary human intercourse.) The real question is whether there are not at least some moral considerations (over and above calculations of national advantage) that ought to enter into policymaking at least some of the time. I believe that evidence of such an international morality is, although not routine, not rare either. We see it in the deliveries of famine relief, the banning of brutal methods of warfare, even in the anguish over the halting intervention in Bosnia-Herzegovina. It is true that any sophomore can construct a self-serving motivation lurking under any apparently altruistic act: "We sign onto warfare conditions to protect our own"; "we feed famine victims in Somalia because storing grain is expensive."

It is hard to rejoin to that sort of response. But also, fortunately, not necessary. For the question, once more, is not the empirical one, whether nations are ever guided *in fact* by moral considerations. The question is whether it makes sense to maintain that nations *ought to* be to some degree moral (and if they are not so now, let us prevail upon them to change their courses).

But even if we suppose there is some such core of international morality, it does not take us very far. A moderate realist will grant that we can identify familiar moral principles which, by easy extension, condemn as evil the torture of prisoners or the rape of civilians. The problem at hand is much tougher because it lacks any well-chartered foundation in domestic moral literature: for example, in urging protection of the biosphere,* to what moral principles can a global moralist refer?

There are, in fact, two distinct types of moral questions relevant to biospheral degradation, each of which raises a fundamental philosophical challenge.

The first task is to identify *a shared international-morality-in-respect-of-the-environment.* To put it simply, if the nations of the world are to cooperate in the reduction of globe-hazarding substances or the protection of species, how are the burdens of those actions to be apportioned?

The second task is no easier: putting aside the conflicts that divide nations from one another: *What are the obligations that humankind, as a whole, owes to the rest of the natural world?***

To illustrate the difference in outlook, the first question could be illustrated thus: If whales are to be protected, have traditional whaling nations any claims for compensation from nations rich in cattle and grain? The second question is the deeper underlying one: Has humankind any duty to whales to begin with?

I will deal with these two questions, in turn.

---

* The biosphere is the thin layer of our planet and its periphery capable of supporting life naturally.

** There is a third, separable question which considerations of space force me to tuck into my treatment of the second, below: the conflict between one nonhuman element in Nature and another. All human action—the decision to plant one crop rather than another, to build a road here rather than there—not only affects the human-Nature balance, but shifts advantage among elements in nonhuman Nature. A decision to reestablish wolves in a park is a decision to kill deer, even to set the stage for their suffering. Identifying, clarifying, and arbitrating this third group of conflicts is within the purview of a fully comprehensive environmental ethic.

Global Justice in the Environmental Context

The questions of the global environment would be hard enough to resolve if the required division of wealth and responsibility were to take place under the most ideal conditions we can imagine: in a world of nations virtually equal in wealth and power, their relations unmuddied by any prior history of hard dealings, in circumstances of such abundance that each appropriator could be imagined to be leaving (in John Locke's phrase) "enough and as good for others." In our time, proposals for divvying up the remaining commons and imposing costs for upkeep of the remainder are met with continuous reminders that if so ideal a garden ever existed, we have long since vacated it.

The most outspoken of the "reminders" have been the underdeveloped countries, most vocally the Group of 77 nations (G-77),[14] said collectively to represent 70 percent of the world's population but only 30 percent of its income. Their mood still finds its most authoritative expression in the U.N. General Assembly's 1974 Declaration on the Establishment of a New International Economic Order (NIEO). NIEO declared the principle of each nation's "full permanent sovereignty . . . over its natural resources and economic activity."[15] The phrase was originally understood as a denunciation of exploitation by others, but today the other side of the coin is considered equally significant and continues to cast a shadow over agreements on biodiversity and forests: the sovereign right of each nation to exploit its own timberland and ecosystems without denunciation or interference by others. At the same time as the LDCs stress their sovereign independence, they insist that the developed states acknowledge a *duty* to reduce the material disparities in wealth.

For many Third World leaders, the emphasis on moral duties and rights is critical. In the words of former President Julius Nyerere of Tanzania: "I am saying, it is not right that the vast majority of the world's population should be forced into the position of beggars, without dignity. . . . The transfer of wealth from rich to poor is a matter of right; it is not an appropriate matter for charity."[16] Casting arguments for assistance in terms of "rights" is not only a sop to Third World pride, it rebukes the rich not to pressure the poor to accept conditions on the transfer of wealth—conditions such as reform of land and population policy, or the elimination of human rights violations.[17]

While division of and responsibility for the commons have not constituted the principal focus of the NIEO (the same issues would have been raised had there been no unapportioned areas, but only bank debts), the new egalitarianism surfaces in whatever forum commons issues are presented. In the LOS negotiations, egalitarian agendas provided a continual source of contention in the context of forming an ownership regime for seabed mineral deposits. The United Nations, well aware that only the wealthy nations have the technological and financial wherewithal to mine, voted that the seabed, ocean floor, and subsoil were part of the "common heritage of mankind,"* not subject to appropriation by anyone or, indeed, even open to exploitation activities pending establishment of an international regime.[18]

The dissension over the seabed wealth that was so vocal in the 1960s and 1970s came to be echoed in space. Here, too, it is the most developed nations that have the wealth and technology to grab the choice orbital "parking spaces" for geosynchronous orbits and frequencies. If traditional international law principles apply, they will get ownership of the choice slots on the same basis that the European naval powers once were awarded title to the Americas and other lands "uninhabited by Christian people"—by becoming the first *humans* to occupy them—"native" populations not counting. Understandably, just as in the oceans context, in space, too, it is the more numerous, less developed nations that seek to construe the distributional question to be one of "equitable access to frequencies . . . a natural resource of humankind."[19]

---

* "Common heritage of mankind" (CHM) (or, as we are now inclined to say, "humankind") has no single accepted operational meaning. The Law of the Sea Convention, presently open for signature and signed by fifty-two nations, undertakes to supply it with detailed content in the oceanic context; the U.S., which regards CHM as implying equal access but not as extending to co-ownership, has refused to sign, principally from reservations over provisions for a world government administration of seabed mineral exploitation.

The same sentiment has surfaced in the debates over the Antarctic. The Malaysians have denounced the treaty system worked out among the nations that discovered and explored the continent as "an agreement between a select group of countries that does not reflect the true feelings of the Members of the United Nations or their just claims. . . . Henceforth all the unclaimed wealth of this earth must be regarded as the common heritage of all the nations."[20]

It should be added that recent estimates have downplayed the near-term commercial value of much of the commons wealth. So far as resources are concerned, an overburden either of ice or sea or, in the case of outer space, sheer distance makes the logistics of extraction and transport intimidating. Anticipated environmental constraints cloud the prospects even further. Beyond that, market demand for most of the target minerals has been restrained or actually depressed. The result might be said to alleviate the urgency of arriving at practically detailed solutions for, say, the seabed, but certainly not to mitigate the symbolic stakes and their significance.

At the same time, the other side of the coin—how to pay for the upkeep of biodiversity, greenhouse blanket, ocean cleanliness, etc.—has only escalated in practical significance. If ozone-depleting agents and GHGs have to be restricted, how should the costs of the efforts be split?

RIGHTS AND CLAIMS OF JUSTICE

Politicians and diplomats are free to talk, as Nyerere did, in terms of absolute rights and duties. But if those demands are to have any moral appeal, one ought to know more about the basis on which they are grounded. The most useful entrée is probably through the conventional distinction (it traces to Aristotle) between corrective justice and distributive justice. Questions of *corrective justice* are precipitated by blameworthy acts. In the familiar context of ordinary interpersonal conflicts, Al has wronged Barbara, either to Al's benefit or to Barbara's injury (or both). Corrective justice deals with what Al must do morally to set the situation aright: for example, to give up his ill-gotten benefit or compensate Barbara for her injuries. *Distributive justice* deals with obligations that arise not from what anyone

247

has done to someone else, but from situational disparities. There is simply a discrepancy between Al's position and Barbara's position (in wealth or power or some other good) that exceeds defensible bounds, and redistribution is said to be in order independent of either side's blame.

The global environmental movement has provoked both kinds of justice claim—on a grand scale.

CORRECTIVE JUSTICE

Most commonly, demands for corrective justice in the international arena take the form of LDCs insisting on some recognition for injuries they allegedly suffered during (and from) colonial domination.[21] Nigerian Chief Moshood Kashmowa Abiola has been pressing for white people in the United States, Europe, and the Middle East to repay Africa for damage done in the slave trade.[22]

In the environmental area, corrective claims of a different sort are arising. Consider India's demand that the developed nations compensate her $2 billion as a precondition of signing the Montreal Protocol on the grounds "that since it is the Western nations that caused the ozone depletion, it is their moral responsibility to transfer technology for CFC substitution."[23] The developed countries are to blame; they should pay what it will cost to clean it up.

One can intuit some force to the Indian claim. But putting a persuasive moral foundation under it is more difficult. To begin with, the advanced nations' development of air conditioners and halonusing fire extinguishers is not quite comparable in blameworthiness with fostering terrorism or waging aggressive war. It is not clear how principles of corrective justice apply to the unintended consequences of acts that are not themselves morally culpable. Kant's general solution to the question of international morality (in *Perpetual Peace*) was: "All actions relating to the right of other men are wrong" if their maxim is not consistent with publicity.[24] Roughly: "If you are a statesperson contemplating a public act that you 'cannot publicly acknowledge . . . without arousing everyone's opposition,' it isn't moral and you shouldn't do it."[25]

Kant's test is not bad as a rough and ready guide, even to this day.

It applies easily to flagrant wrongs, such as torturing prisoners and, more in our context, the dumping of precariously packaged radio-active or toxic wastes in inhabited areas. The test also has the virtue of an application that expands with the growth of environmental consciousness; heads of state are not as prepared as they once were to openly stand behind drift-netting and elephant-poaching.* But it is of less help where we need guidance most: in regard to the broad range of more subtle or at least more ambiguous insults to the environment, such as deforestation in order to provide needed farmland. After all, heads of state have commonly been well prepared to stand behind many environment-affecting actions on the basis not just of naked sovereign prerogative, but of higher moral need. And in the specific illustration of India and CFC congestion, it is fairly clear that, given our collective ignorance about the perils, no head of state or anyone else would have disavowed the uses of CFCs during the buildup from the 1930s through the 1980s.

There is another argument that countries such as India might raise in favor of their position—if they would endure appealing to an English philosopher. It is, after all, John Locke to whom the developed world naturally looks for a moral defense to the rule that ownership of things held in the common (such as the fish of the sea, the acorns of the forest) is awarded to the "first occupant." Locke maintained, however, a less well known proviso to that principle—that the first appropriator's title holds only as long as the appropriator leaves "enough and as good for others."[26] In other words, if there is an abundance of fish, so that the fisher leaves for others all that the others can use, there is no quarrel with his taking fish from the common pool.

But how does Locke apply in our real world, in which consumption is so often rival (more for me means less for you)? It is open to

---

* Ironically, it is not likely that Kant, the staunchest opponent of moral relativism, would have ascribed true flexibility to this test, or considered flexibility a virtue. He might have acknowledged that as societies evolve different levels of knowledge, moral sensitivity and so on will issue in different moral judgments—each of which will be comprehensible as an application of the same unvarying (if general) moral law.

India to argue that the proviso dominates, and that over the past two centuries the industrial world has been congesting the atmosphere—"taking" its safe-range absorptive capacities—without leaving, in Locke's phrase, "enough and as good for others." Along these lines, the Third World might draw from Lockean thinking a basis for corrective justice claims against the developed countries.

But even if we credit the strength of this line of argument, it runs into peculiar problems in application to nations. It is hard enough to say how corrective justice applies among ordinary mortals in ordinary family and neighborhood relations. But in this context the questions multiply. Even if we assume that nineteenth-century American slave traders injured nineteenth-century Africans in the slave trade, it is far less clear that modern-day (white) Americans are obliged to repay modern-day (black) Africans—much less how any payment would be measured. If contemporary Britishers are under a moral obligation to make reparations to modern-day Indians for injuries done in colonial rule (even though colonialization was in keeping with the then accepted norms of the world order), may the British offset India's claims by the value of infrastructure received by India, including a common language, nationwide system of law and administration, and so on? Can an activity as well motivated (but ultimately harmful) as the manufacture of air conditioning units give rise to claims that carry from one generation to the next?

I emerge from these considerations unpersuaded that there is any body of universal corrective justice with a morality detailed and thick enough to govern global conflicts on its own terms; there may well be—I believe there is—a substratum of international morality, but rather than being self-executing, it requires continuous appeal to conventional norms embodied in legal rules and treaties. In other words, nations are obliged to deal justly with one another, but in any concrete circumstances we cannot specify the justness of their expectations except against a background of preexisting rules—rules of trade, of warfare, even, more recently, of the environment. The better formulated of these rules embed their own corrective justice formulas, specifying under what circumstances, and in what amounts and ways, reparation will be owing. But absent such rules, the filaments that relate nations across the globe appear too thin, and the expectations are too nebulous or scattered by conflicting moral codes for us to construct extensive and detailed guidance from the raw materials of any moral theory.

DISTRIBUTIVE JUSTICE

The distributive justice claims are appearing in the environmental context even more ubiquitously than those for corrective justice. That is because most of the controversy today focuses less on recrimination for past wrongs than on preventive measures to reduce degradation in the future. Unfortunately, the conflicts that arise over burden-sharing are particularly hard to mediate. The Law of the Sea negotiators had a tough enough time figuring out how to divide the anticipated (and one would now say, exaggerated) new wealth that was to be drawn from the oceans. In the pollution-oriented treaties, the negotiators are faced with offering everyone a smaller cut of the pie.

Most would agree that to cut the pie justly is to do so fairly. But what is "fair" in this context? One view of fairness is that it demands *equality of effort* to reduce pollution. But do we have equality of effort when each nation has expanded the same sum on abatement efforts: $1 billion per nation? Or does fairness require the same percentage of each nation's GNP? Or installation by each nation of its "best available" technology?

On the other hand, the equality could be understood as an equality not of effort but *of outcome*. But outcome is ambiguous also. Some would say that the fair outcome is an equal reduction in units of emissions: every nation cuts back 100,000 tons. But if that is the process, the nations with the highest historical baselines will continue to outpollute the less developed nations—and maintain, "unfairly," their economic edge. Those who press for fair outcomes might therefore choose to aim for an all-things-considered outcome: each nation should put in so much effort at pollution control until we all have not an equality of emissions, but an equality of wealth or of opportunity or at least of some fundamental baseline index that assures a floor of adequate food, fuel, clothing, and the like.

251

The difficulties of sorting through these competing standards of "fairness" to find the morally right one is frustrating. And that frustration, in turn, provides a major boost for the various market solutions, which advertise the "unseen hand" of the market as rescuing the human mind from hard choices. Various schemes for marketable pollution allowances (chapter 6) are a prime example. But most of the market-trading literature deals with the techniques and benefits of trading. It assumes someone else has provided an answer to the threshold question: How shall we assign the original entitlements, the starting point from which the trading begins to operate? Is the right to pollute a personal right of each member of the human species, to be handed out per capita? Or is it a geopolitical right, to be allocated, like a vote in the U.N. General Assembly, evenly among nations? The snag with the first alternative is that it undermines incentives to control population* and is theoretically obscure, anyway: Should each person on earth have a pro rata right to the globe's reserves of oil, fish, timber, and farmland? The problem with the second alternative is that to divide pollution entitlements evenly among nations would give the tiniest nation the right to pollute as much as the largest. Do we take the status quo as a starting point, so the rich can fortify the advantage they have gained over the poor, at the poor's expense? And so on.

These are not questions of economics, but, unavoidably, of ethics. Indeed, in any social philosophy, the just distribution of entitlements—of power, wealth, office, opportunity—is among the hardest, most foundational issues. The stakes are all the more momentous when applied on a global scale, but all the more problem-ridden too. Moral philosophers have their hands full warding off skeptics who charge that ordinary moral discourse is at bottom "meaningless." Philosophers who want to apply moral predicates on the international plane—to say that some international acts are "good" and others

---

* This problem might be finessed by assigning each nation an emissions quota based on its population in the start-up year, with no credit for subsequent increases; of course, this strategy wanders away from the original underlying rationale, that each individual on earth should have an inherent equal right to pollute.

"evil"—face additional challenge from the realists, above, who maintain that even if moral discourse is intelligible generally, acts of state are beyond its purview. And then there is another tier of challengers, "reductionists" who mount not a moral but an even more fundamental metaphysical assault on the evaluation of national conduct. Persuaded that "states do not act, only people do," these critics never even reach the question whether the actions of states lie beyond good and evil, since there are in reality no *state* actions to evaluate. The notion of a state acting, rightly or wrongly, is simply incoherent. Only individuals act and are acted upon, have interests to advance or frustrate, can be praised and blamed.*

THE CONTEMPORARY PHILOSOPHICAL BACKGROUND

Defending the global environment will cost huge sums of money. The burdens, particularly of maintaining the commons areas, have to be somehow distributed. It is hoped that they can be apportioned according to some notion of justice. But what does justice demand? In what circumstances does it apply? And between whom—or what—do claims of transglobal justice run—nations or people?

The contemporary reference point for these questions has become

---

* This last issue sounds metaphysical and abstract, but the stakes are high. Ought we to embargo trade with Iraq, causing suffering to its people, on account of the actions of its leader? The classic utilitarian response is reductionist, to disregard states, and look only to the outcome in individual welfares. (The embargo will be wrong if and only if it can be shown that more people will suffer in the long run from not embargoing than from embargoing.) Others, including myself, are disinclined to dismiss states as transparent in all cases and may inject into some discussions of embargo or aid respecting Nation N that N's failure to abide by a U.N. resolution or control population is a national failing of N—of its government and culture (like the loss by a team, which cannot be redescribed simply as the loss by each player). In the moral plane that relates nations to nations, N's failing can be viewed as affecting analysis of our obligations, if any, that tie individuals across the globe. On the other hand, I do not believe that this consideration nullifies the obligation to provide famine relief, which is perhaps less to be treated as a question of justice than of humanity. This is a very complex area, which I have examined more fully elsewhere. See Christopher D. Stone, *Earth and Other Ethics* (New York: Harper and Row, 1987), chapter 16.

John Rawls's monumental *A Theory of Justice*. But in our area of inquiry Rawls's ambitions were uncharacteristically bridled. He drew a sharp line between the principles of justice that prevail among persons within a society, about which he had much to say, and "justice between states," which he considered to be much thinner and less pregnant. The entire subject of international justice, which I believe will be the great philosophical issue of the 1990s, Rawls touched upon in 1971 only indirectly in the course of illustrating conscientious refusals of citizens provoked by differences over foreign affairs; the entire treatment of international justice was thus disposed of in less than three pages[27]—too obscurely to rise to his indexer's attention.

The short shrift given global justice by Rawls flows in part from the times and in part from his conception of how principles of justice are established. He asks us to imagine ourselves as part of a hypothetical negotiation among all members of society at the time they came together into a social union. All the negotiants are assumed to be predominantly interested in their own welfare, but are to imagine themselves hammering out the society's ground rules unaware of the particulars of how their lives will unfold. No one knows whether he or she will turn out to be a man or a woman, poor or rich, talented oboist or mentally retarded. The idea of laying down the framework from behind such a "veil of ignorance" is that, not knowing our gender, race, and so forth, we will vote for rules that are gender and race neutral—"fair," in the sense of not narrowly self-serving, and therefore "just."

Rawls maintains that, among other choices, the negotiants would agree on a set of rules reckoned to produce an equal distribution of primary goods.* But there is a caveat that Rawls calls the Difference

---

* Rather than to express the principle in terms of welfare, as he might well have done, Rawls chose (1) to speak of an index of primary goods including rights, liberty, self-respect, wealth and income (*Theory of Justice*, 21), and (2) to conclude that those in the original position would maximize the least well-off negotiator's basket of those goods (implicitly assuming, but not explaining, a high degree of risk-aversity). John Harsanyi seems more persuasive, that under a veil of ignorance in which individuals do not know their abilities, tastes, social position, etc., they would choose the regime

Principle. Unequal distributions are legitimate to the extent that the rules that permitted the inequalities to evolve make the least well off better off than they would have been without the rules. In other words, rules and institutions that make everyone better off are acceptable, even if some people emerge with more pie than others, as long as they maximize the welfare of the minimally well off.[*]

The implications for international society could be weighty. To defend the rules of international trade or the concentration of carbon-belching industry in the industrialized North, an American might argue that although the underlying rules and practices that countenance them (those of international trade, banking, etc.) leave great disparities in wealth, they nonetheless operate to make even the least fortunate in Bangladesh better off than they would be otherwise—without those rules. That, at least, is how the argument goes. But can we really say with a straight face that if we were to surrender bits and pieces of our high technology opulence, the bottom would fall out of their jute and copper markets? The NIEO evidently thinks otherwise, that the practices which sustain existing disparities are simply stacked for the North, without any absolution measurable in long-term benefits to the least well off (or possibly even in the average welfare globally).

---

that maximized their expected average or perhaps even their total utilities. John Harsanyi, "Cardinal Welfare, Individualistic Ethics, and Interpersonal Comparisons of Utility," *Journal of Political Economy* 63 (1955): 309–21. For a general discussion, see Amartya K. Sen, *Collective Choice and Social Welfare* (San Francisco: Holden-Day, 1970), 131–51.

[*] A rule allowing unrestricted incomes—for example, there is no legal limit to what a corporation can pay its executives—certainly aggravates disparities in wealth in the United States. The practice might be thought therefore prima facie unjust. But supporters of the law would argue, appealing to Rawls, that uncapping salaries provides incentives for executives to work harder; and that enough of the benefit of their added work "trickles down" to those at the bottom of the social ladder so that those at the bottom, though much less well off than those at the top, are nonetheless better off than they would be if those disparities and incentives were not permitted. Harsanyi, note above, would recast the same argument in terms of expected average welfare.

Who is right? If one is willing to accept that the international order is based upon sheer power, then the question of "right" does not enter and it is simply going to come down to our armies versus their terrorists. But for those who take morals seriously (who want the international order to be built on something beyond sheer power plays), the questions of justice are vital: Need the developed world defend existing practices by justifying them with some sort of Difference Principle defense?

Ironically, Rawls himself did not think so. He conceived his "veil-of-ignorance" thought experiment, above, to apply only to persons engaged in "a cooperative venture for mutual advantage" in a "self-contained community."[28] Rawls felt that the relations among people within each nation met the standards of cooperative interchange, but that the quality and extent of interaction among nations themselves did not.*

In consequence, on Rawls's view the principles of justice that prevailed internally stopped at national boundaries. While there were principles of international justice, they were special and limited. The rules are those we would get if, after the members of each nation had worked out their inward-looking justice principles, a second convocation of national leaders were held. The national leaders would have to imagine themselves behind their own veil of national ignorance, knowing that they represent nations (in principle) but unaware of whether they represented a big country or small, a naval power or an LDC.[29] Rawls imagined that the results of this convention would produce "no surprises"; "the principles chosen would be . . . familiar ones" such as can be found in standard international law texts, such as norms touching nonaggression, the sanctity of treaties, the laws of warfare, and so on.[30]

---

* Several commentators would argue that to globalize justice would simply "overstrain our commitments." Rawls, in a subsequent work expanding slightly on *A Theory of Justice*, reaffirms his conclusion that the difference principle that is championed domestically "is not suitable for justice between states," but the bases of his judgment are still not spelled out in detail. John Rawls, *Justice as Fairness: A Brief Restatement* (Cambridge, Mass.: Unpublished, 1989), 151–53.

The passage regarding international justice commands so little space in *A Theory of Justice* that it was generally neglected by most of Rawls's commentators. But a few critics, catching the significance, lamented the fact that Rawls has nothing to say about population control, resource redistribution, division of the commons wealth, or the global environment.[31] One of the principal critics was Brian Barry, who took Rawls to task for not pursuing the implication of a veil of ignorance to its logical conclusion and, in effect, globalizing the difference principle.

To dramatize what such a globalization would look like, Barry asks each of us to suppose "that you were a random embryo," and to ask ourselves:

> What kind of world would you prefer? One, like the present one, which gives you about a fifty-fifty chance of being born in a country with widespread malnutrition and a high infant mortality rate and about a one in four chance of being born in a rich country, or a world in which the gap between the best and the worst has been reduced? Surely it would be rational to opt for the second kind of world; and this conclusion is reinforced if we accept Rawls's view that an element in rationality is playing safe when taking big decisions.[32]

The approach is provocative. If we did not know how things would come out—whether we would wind up flourishing in the United States or (with much higher probability) struggling in the Third World, what sorts of rules would we vote for?

To start with one of the less world-shaking ramifications, but one relevant to the global environment, my guess is that most people would *not* vote to institutionalize the widely advocated entitlement of each nation to its pro rata share of allowable carbon emissions based on population. If there are (as there would have to be) appreciable costs to trading, such a rule makes no sense. To see why, recall for a moment that farmland is a valuable global resource, too. Imagine a proposed rule that would allocate rights-to-farm among nations. Canada and the United States would be prohibited from harvesting more grain per capita than Kuwait and Bahrain. (Another rule would prohibit Kuwait and Bahrain from producing more than their per capita share of oil.) Assuming that there were costs to arranging the purchase of Kuwait and Bahrain's rights to grow crops and the sale to them of other countries' rights to produce oil, such a rule would yield less food and oil than is produced currently—hence, less real wealth for the world. There is no reason to suppose that it would make the least well off in the world any better off—if that is the test.

Now, the same thing seems true of national pro rata carbon permits. Just as the global negotiants would know (for their "veil of ignorance" does not deny them access to a good world atlas) that some nations would have comparative advantages in farming and others in oil production, so, too, some nations, on account of access to harbors and trade routes and early forms of energy, would develop into the most efficient industrial centers. Imposing special limits on each nation's "fair share" of fossil fuel use would not maximize global wealth within the constraints of climate safety. It would make no more sense than imposing national quotas on farm or oil production.*

But that does not mean that the negotiants would endorse the world as it is. It is in fact hard to imagine that the original international contractors would not agree upon some redistributive principles.

Charles Beitz, one of the most thoughtful of Rawls's commentators to take up this question, begins by distinguishing "two elements that contribute to the material advancement of societies. One is human cooperative activity itself, which can be thought of as the human component of material advancement. The other is what Sidgwick called 'the utilities derived from any portion of the earth's surface,' the natural component."[33] Then, focusing on the nation's natural resource endowment, Beitz likens it to an individual's talent, that is, "arbitrary from a moral point of view." Hence he regards a nation's natural resources as readily and fully subject to redistribution on the view that one's "natural capacities" are a part of the self, "in the devel-

---

* Granted, making the quotas tradable would mitigate the nonsense, but the trading is hardly costless.

opment of which a person might take a special kind of pride."[34] From this, Beitz argues that even if we were to go along with the assumption that nations are self-sufficient, principles of resource redistribution from resource-rich to resource-poor nations (a correction for this morally arbitrary lottery) is surely "a subject that would be on the minds of the parties to the international original position."[35]

I find this analysis forceful but not convincing, however. Why should natural resources or the benefits derived from them be subject to redistribution,[36] rather than to focus redistribution on other indices of inequality such as total wealth or primary goods, or (Sen's suggestion) primary powers?[37] One would think all of these can vary from country to country in ways that are no less morally arbitrary from the perspective of the "random embryo" and probably more directly influential on the life she will lead.[38] Further, by choosing natural resources (rather than wealth) as providing the pot for redistribution, Beitz's justice has the odd implication of underwriting a transfer payment from resource-affluent Zaire to resource-poor Japan.

Indeed, the emphasis on resources (as distinct from the human component of a nation's wealth) seems increasingly incoherent the farther civilization has progressed. That is, at the beginning of civilization, the resources each group found under its feet might have been, as Beitz suggests, a matter of moral luck, utterly arbitrary. But once, crossing straits and mountains, people moved to and from resources, learning to exploit some and wasting or overlooking others, untangling wealth into two components is not just hard; it may not be intelligible. Consider an LDC's uranium or platinum or even diamonds. Are we to assign their values to nature—to the worth of the minerals in the ground?—or to the highly organized human activity of the advanced scientific and opulent nations, without which the minerals would lack any appreciable commercial market, would be at best pretty baubles? What would Saudi oil be worth, were it not for the invention of the internal combustion engine and the freeways of Los Angeles?

I doubt therefore that a justice conference would yield a rule of resource redistribution. My guess is that two sorts of rules—perhaps in the alternative or in some combination—would emerge. The first would be some unconstraining of boundaries. If we did not know where we would be born but were aware that conditions could vary radically, we would probably consider open (or more easily permeable) borders a condition of a just world order. I say probably, because it is not a sure thing. Each of us would have to weigh the prospect that we might find ourselves living in a community we found pleasant, but which would be, under an open-borders rule, subject to being disturbed if not overrun by others.

The second sort of rule would look beyond indirect indices of wealth such as carbon use or resource endowment as a reference for welfare–leveling and—let's face it—open up wealth itself. A wealth-regarding redistribution would not entail a global wealth-leveling tax, however. There are two reasons. First is the traditional argument that a highly progressive tax (one that sought to level post-tax welfare) would erode incentives and therefore be rejected on the grounds that it would imperil wealth on average.*

The other constraint on a wealth-leveling tax is more subtle—and has more general application for us. To understand it, let me go back and question Barry's position that, to create the original (international) position, we imagine ourselves in the position of a fetus not knowing where it will be born. We might better imagine ourselves coming together as normal adults, not knowing where we will have been born, but knowing that, in principle, we will have evolved mature tastes, settled into an established pattern of living, and so on. We each have to weigh the risks of a sudden radical restructuring in midlife, perhaps upward, perhaps downward, depending on (what is unknown to us) the circumstances in which we will find ourselves, compared with the rest of the world. I raise this alternative because there is reason to believe that across a broad margin of wealth the

---

* As indicated above, Rawls's test suggests that a redistribution would seek to bring the least fortunate person's mix of primary goods toward the mean level; a simplifying assumption is that a tax would move people's wealth toward the mean.

risks of increases and decreases of equal magnitude are not valued symmetrically.* *Losing* a buck is somehow worse.

This asymmetry has two interesting implications for us. First, it suggests that if Rawls's justice principles are to be applied (contra Rawls) transglobally, the original contractors, coming into the negotiation rationally worried about the prospect that they might be caught up in a radical downward shuffling of wealth, would be likely to temper the leveling potential of a Difference Principle with side rules that insure against highly dislocating changes.[39] I imagine that those rules would look much like the rules legal systems across the world embody in periods of limitation. The state cannot try a person for theft, or you cannot eject a trespasser who moved onto your land, twenty years later. All such rules recognize the value we attach to leaving expectations, even "wrongful" ones, undisturbed.

Indeed, the neo-Rawlsian negotiants might well emerge with some "entitlements" which, while perhaps not quite so unyielding as his nemesis Robert Nozick's,[40] would nonetheless impede changes in established practices that have operated perhaps "unjustly" over time, but on which people have relied. For example, the advanced nations might justifiably argue something like this. We are not saying that we necessarily have a right to continue polluting as much as we have been. But just as a person who, after walking across a corner of his neighbor's lawn every day for twenty years to get to his garage gains a prescriptive easement in the pathway (a right to continue the use), so, too, some recognition has to be given to our settled expectations and the high costs of disturbing existing patterns of behavior. What I am suggesting is that the notion of a partial easement to pollute is not at all far-fetched. It may be exactly what original nego-

---

* Daniel Kahneman and Arnold Tversky have demonstrated this asymmetry empirically ("The Psychology of Preferences," *Scientific American* 246 [1982]: 160–73), but the notion is one Mark Sagoff, examining the disparity between prices we bid and ask for the same things, traces back to Hume's notion of hysterias, and relates to our resentment of being asked to pay for what one believes one already owns or has a right to. Mark Sagoff, "Ethics and Economics in Environmental Law," in Tom Regan, ed., *Earthbound* (New York: Random House, 1984), 155.

tiants, operating from behind a veil of ignorance, would have agreed to as "just."

Second, I think that the same analysis reinforces my argument for a Global Commons Trust Fund (chapter 8). Imagine that the original negotiants foresee that some redirecting of wealth may be required. They know, too, that they will be risk-averse to a sharp decline in life-style should a course correction have to be made in their generation. It seems to me that before they would agree to a rule exposing established wealth and income streams to a global tax, they would expose the "new" unallocated wealth of the commons areas first. To illustrate, if $1 billion had suddenly to be provided to clean up oceanic radioactive wastes, my guess is that they would prefer the sum to come (to the extent available) from auction of satellite slots than from a surcharge on national taxes.*

Before leaving this discussion of international justice, there is one last idea I want to toss out—not as a final solution, but as a candidate for further discussion. Essentially, if justice calls for a virtual pooling of the world's wealth (even allowing for some mitigating side rules) it demands too much to be acceptable outside the classroom or pulpit. But international justice need not demand such a far-reaching reshuffling of wealth. Let us go back and follow Rawls in grounding justice in the gains of social cooperation (rather than to emphasize the social contract behind a veil of ignorance). One could understand this to mean that once Nations A, B, and C reach a particular threshold or quality of social interrelationship, from there on justice demands a thorough pooling of A, B, and C's total resources or other assets: if C flags, A and B have to dig into their pockets however far is required to satisfy the principle of maximizing the welfare of the least well off.

But we could also understand talk about the social gains from

---

* Once more, this is not to deny that if fees are collected for uses of the commons areas, someone's welfare, for example, that of stockholders in satellite corporations who no longer get slots for free, will go down. Tracing through such charges to identify who ultimately bears their incidence is no mean task.

cooperation to imply something else. A, B, and C, independent in some undertakings and interdependent in others, might contribute to the common fund subject to redistribution only the gains derived from their social cooperation, withholding from one another's claims any wealth that is not the fruit of the interdependency. In other words, on this conception, justice would function somewhat like a commercial limited partnership.

To illustrate, suppose that A, B, and C are three neighboring coastal states, each with its characteristic wetlands, shorelines, fishing tastes and traditions, and so on. As a physical matter, without cooperation A *could* take from its territorial waters 10 tons of fish in the current year, B, 20, and C, 30. Of course, if any one of them seizes its maximum, the share remaining for all will decline over time in some indeterminate way. Contrariwise, if each takes less than the maximum it can land in the present year, the share available to each will gradually grow. Let us suppose that there are several far-reaching options that will, over time, stabilize the catch at up to 100 tons per year—a gain of 40 tons over the 60 that the free-for-all would yield. Achieving this maximum would require a certain amount of cooperation, however, as to seasons, seine size, maintenance of areas for spawning, and so on.

While such measures could increase the sustainable yield, because the species migrate and any arrangement will shift the ecological balance, no one can be sure exactly how the gains from any agreement will be distributed off each nation's coast. All they know is that the optimal management and harvest of the region will benefit each, but not each equally. In these circumstances a treaty that attempts to fix tonnage allocations in advance may be unfeasible.

Instead, it is tempting to follow Rawls and suppose that each nation will want (and ought justly to receive) an equal voice in the distribution of roles, offices, responsibilities, and opportunities of the fisheries management. Moreover, while they recognize the likelihood that unequal benefits may accrue over time, none should be allowed—or can morally demand—to extract from their mutual efforts a benefit not required by features of the arrangement that operate to the advantage of the nation least advantaged. In essence, this is to recognize some sort of Rawlsian "difference principle" as a requirement of justice.*

A closer examination than I can engage in here, drawing on coalition analysis and game theory, would appear to suggest that Rawls's own notion of a difference principle is more underdetermined than conventionally observed.** But there is no need to over-worry the fine details. The point to take away is simply that while there is a range of "just" solutions within Rawls's constraints, the "kitty" that A, B, and C subject to the justice principles could well be the gains (variously measured) that derive to each from the socially cooperative arrangement. If the catch at its most aggressively—even strategi-

* We have enough to do here without figuring what form of difference principle might emerge, whether Rawls's maximin (maximize the well-being of the least well off) or just plain averaging the sum of everyone's ex ante expected utilities. The point is that if we had to supply a term to the arrangement from morals—interpreting an ambiguity in the A-B-C treaty, for example—the claims of the least advantaged entrepreneur might well be limited to the resources committed to and arising *from the scope of the common enterprise.*

** To suggest the direction the analysis would take, assume that A, B, and C only care about the size of the harvest (so that we can ignore the costs of fishing). Assume further that under the most aggressively competitive circumstances the catch would fall to an average of 20 tons per year—10 for A, 20 for B, 30 for C. Under cooperation the total could rise to 100 tons. Has justice nothing to say beyond the command that A's share not fall below 10 tons per year? There are at least three intuitively appealing bases for dividing the cooperative gains of 40 tons per year, all of which are consistent with the floor. First, there could be an equal sharing of the surplus, each to receive an additional 13 tons; second, the sharing could be proportionate to their respective baseline harvests, so that A, B, and C received an additional 7, 13, and 20 tons respectively; third, complete equality, with each taking 33.33 tons. Each division appears to meet the spirit of Rawls's minimum justice criteria, in that no party is less well off under the agreement than outside it.

The analysis is further complicated the more critically one examines the notion of baseline. Both the Nash and other solutions to the bargaining game require agreement on the appropriate no-agreement point. Should the point be the amount each party would receive if it acted in its own self-interest, or should it be the amount each party would get if it were determined to underscore the harm it could do the others, the so-called "threat point"? Perhaps A would take 16 tons even if it were uneconomical to do so, just to improve its bargaining hand.

cally—competitive circumstances would fall to 60 tons and under cooperation rise to 100 tons, then perhaps 40 tons are subject to redistribution (depending on time frame and adoption of the ordinary self-interest or the threat-point baseline), not necessarily the full 100 tons, much less *A*, *B*, and *C*'s total national resources.

Observe that the same sort of "justice" analysis could be introduced into negotiations over climate change, ozone shield, and other problems. One guiding ideal could be to distribute burdens not in proportion to national wealth as such, but to the benefits each nation could be expected to derive from the proposal at issue.*

It is instructive that thus far none of the major international environmental accords has attempted even to approximate any such theoretically fine-tuned solution to the distributive justice tensions. As a response to wealth differentials, the Montreal ozone negotiators simply deferred the compliance schedule for less developed nations, defined for these purposes as countries whose annual level of consumption of controlled substances is less than 0.3 kilograms per capita. There is also a proviso allowing an increase of up to 10 percent for less developed countries that need that amount to satisfy "basic domestic needs," as well as provisos for some "trading" of rights for nations at the lower end of the development scale.

Treaty-makers are not academics. (We would almost certainly not have had the ozone accords if they were.) Their job is entirely practical, and when distributive claims are aired the negotiators are inclined to forge compromises in rough and ready and largely intuitive approximations of what "justice" might require—particularly if there is no clear and persuasive argument that a practically achievable agreement is palpably *unjust*. And indeed, if a draft convention can attain the requisite consensus and is workable, there is no need to trouble unduly over its philosophical foundations. But considering that the future will see increasing efforts to form international regimes, often with profound effects on nonparticipants, surely more thought should be given to the underlying moral issues that such

---

* This approximates but is not exactly equivalent to what each nation would be willing to pay for the proposed plan on a take-it-or-leave-it basis.

arrangements raise. Principles of global justice—not merely as they touch the environment but as they inform our actions regarding trade, armaments, human rights, treatment of ethnic minorities, and more—are going to count among the most important topics of the new world.

Of Humankind and Gnats

We have left the hardest for last: the construction of a true environmental ethic. Put aside for a moment the issues that divide nations. How can we arbitrate the conflicts between humankind, on the one hand, and the rest of the natural environment, on the other? Environmentalists incline to downplay the sharpness of the conflict, pointing out that if we foul our nests, strip the forests, and eradicate species, we ourselves suffer, along with Nature.* Indeed, many of the main items on the conservationists' agenda, such as protecting the ozone layer and arresting deforestation, need no philosophically exotic justification. We are far from doing what ought to be done, just from traditional considerations of human health and welfare. Similarly with respect to conserving a wide spectrum of plant and animal species: doing so can stand on the grounds of humankind's benefit, since genetic diversity helps insure the resilience of the food chain against the stress of pests, climatic changes, and man-made environmental toxics.

What is the call, then, for a distinctly environmental ethic? First, there is a wide—and I think widening—sentiment that to argue public policy on purely homocentric grounds is unenlightened, even arrogant. Humankind is maturing beyond its need to presume itself the universe's hub. A nonhomocentric foundation for social choice, even if it should support many of the same choices as would ordinary utilitarianism or some other familiar theory, would be more in line with our growing understanding of the grandness and interrelationships of all the natural world.

---

* The tension here is much like the tension we noted earlier, between the environment and development. There is indeed some overlap to identify and capitalize upon; but there is a tension as well that cannot be ignored.

But the search for an environmental ethic is not just the search for a new rationale to justify the same old ways of life. Humankind's interests and our obligations to Nature may in some sense part ways.[*] Ever since Aldo Leopold, conservationists have been fudging the congruity between conservationism and human welfare. Leopold boldly announced the need for a new "land ethic," but immediately wavered to sound like an enlightened utilitarian, justifying the ethic on the grounds that treating the earth well—understood in some human-independent way—will inevitably redound to humankind's advantage.

Of course, as far as it goes, some such wavering is perfectly understandable. Ravage nature and we hurt ourselves. But the truth is that not all the conservationists' goals, in their fullest reach, can be persuasively or ingenuously defended by appealing to the greatest good of the greatest number of humans. For example, utilitarianism—the greatest happiness of the greatest number—is surely an awkward basis on which to defend the perpetuation of wilderness areas, which would frankly provide more people more enjoyment if they were opened up to wider access, even to Disneyland development. On the same grounds, if human utility is the judge, then Third World leaders and corporate executives who want to open the Antarctic to oil and mineral development have the best of the argument. And again: there is, undoubtedly, potential human benefit to be realized from rain-forest ecosystems, solid benefits that warrant preserving appreciable forest cover on the basis of fairly straightforward utility considerations. Over some margin, the value of the land to humans as a forest-covered genetic storehouse dominates its value to humans as paved–over habitat. But I am not so persuaded that the degree of conservation that many biologists and environmentalists are—in my view, rightly—seeking can really be defended by an appeal to any utilitarian calculus, even one that gives a liberally heavy weighting to "enjoyment" of Nature. On the other hand, human benefit is not really their genuine motivation or justification.

[*] Inevitably a complex sense, inasmuch as while humans take an interest in Nature, and some things in Nature—whales for example—have preferences, it is clear that Nature takes no interest in itself.

Indeed, to me there is always something pathetic about appeals on behalf of Nature that have to fall back on inflated claims of derived benefit to homo sapiens; it is like saying we should not kick dogs *because it is painful to watch*; or *because it will come back to haunt us in increased violence among humans*. I suspect that people who talk that way are really thinking along lines that are at once more sensitive and more noble, but we lack an ethical framework in which to express and defend what we would really like to say.

WHAT DOES AN ENVIRONMENTAL ETHIC AIM FOR?

The task of constructing a genuine environmental ethic is nothing less crucial than to provide that missing framework. To appreciate why it is needed, and how it would operate in thought, imagine a proposed project for which no one candidly claims warrant in terms of present or even future human welfare, as we ordinarily understand it. The decision to reintroduce wolves into Yellowstone (chapter 7) is a good example to pursue. Humans have not acquired a taste for wolf meat, and, even worse for the project's proponents, wolves destroy things humans raise to eat themselves, such as cattle, and other things people like to shoot and mount on their walls, such as moose.

We saw in chapter 7 an ingenious (but partial) solution to the problem. Conservationists who want to foster the return of the wolves have established a fund to indemnify ranchers for any cattle the wolves destroy. If the fund the conservationists can raise continues to prove ample to satisfy the ranchers, then there is no further conflict—and no need to call upon any special, peculiar environmental ethic for further guidance. We will just have one group of humans putting up the cash to reach a satisfactory, "efficient" agreement with another group of humans—an agreement in which the wolves are what lawyers call third-party beneficiaries.

The issue becomes tougher if the money that the conservationists are able to raise proves inadequate to cover the costs of reestablishing the wolves. In those circumstances the verdict of ordinary homocentric morality is clear: from a preference utilitarian point of view, humans value the lost steak over the added wolves—and so the wolves must go.

But there is no reason why environmentalists should have to accept that judgment. The modified market solution having brought us as far as conventional economic analysis can travel,[41] the issue could be advanced to the political arena. Should the return of the wolves be subsidized from the public purse, on the grounds of correcting not a "market failure" but a moral one? To pursue this line of thought is to plunge into deep political theory. If one adopts a passive conception of legislative power, in which the representatives restrict themselves to mirroring and making the best "trades" to advance what they presume to be the well-formed desires of their constituents, then there is no justification for the legislators to "overrule." The modified market test already created a strong presumption of what the voters wanted: more cattle, fewer wolves. But Cass Sunstein contrasts this horse-trading, "pluralist conception" of democracy with a more independent and deliberative "Madisonian conception" under which representatives are licensed, indeed, encouraged, to go beyond the question of how to satisfy already well-formed desires and venture into the discovery and evolution of new ones.[42] In other words, the selection of preferences is among the objects of the government process.[43] That view of community underwrites a continuous, representative-led evolution of values including a shift in the balance between the well-being of humans and the flourishing of their environment. The context of the shift would be the legislative discussion of putting a monetary value on the return of the wolves, as it once did on their extinction.* The value of the wolves

---

* The state of Minnesota has in fact included in its wolf restoration program state reimbursement to stockraisers for wolf-destroyed cattle from public funds. My understanding is that overall the program is considered successful. One point of discord is that proof of loss, ordinarily a carcass, is a requisite of payment, and some victims of wolf predation are simply dragged off into the woods, never to be found. Wolves caught predating on cattle (apparently some wolves incline to prey on domestic flocks, others keep their distance) are trapped and put to sleep partly as means of appeasing the ranchers.

is certainly not going to emerge as infinite—that is, as over-riding every conflicting human desire.[44] On the other hand, the legislature need not accept as definitive the prevailing value their electors put on wolves versus cattle, based on their willingness to pay enough into the wolf fund to outbid what the market will pay for lost steak.

The debate would be expected to grope with the rights of the wolves to share in Nature; the significance of the wolves having been here first; the question whether they have not as much right to eat cattle as humankind. Indeed, perhaps the wolves have more right than humankind to the meat the land can support, lacking, as they do, our capacity to make moral choices and our wide-ranging alternatives to select other equally beneficial diets. Some legislator might even quote from Leibniz: "It is certain that God sets greater store by a man than a lion; nonetheless, it can hardly be said with certainty that God prefers . . . a single man to the whole of lion-kind."[45]

Of course, the problem arises when another legislator denies that wolves have rights (or, more likely: "Come back and see me when wolves and lions get the vote"). How do we arbitrate between them? What can either legislator say next to advance his or her point?

It is that exchange which an environmental ethic aims to carry further. The need for such an expanded ethic cuts in wherever, for example, the probable utility benefits of sparing a wilderness area or species, even when the consequences to mankind are intelligently and liberally estimated, do *not* exceed the costs of foregone timber, crops, and living space. In those circumstances, the person favoring preservation, having exhausted appeal to his neighbors' enlightened self-interest, has to invoke *something else* as a warrant for the sacrifice. But that is the nub of the controversy. The skeptic will maintain not merely that it is unwarranted for us to make sacrifices on behalf of species, or even, less plausibly, of inanimate natural objects such as rivers and wilderness areas. More severely, the skeptic will maintain that we cannot even make such a pro-Nature argument *intelligibly*.

270

THE ARGUMENTS AGAINST AN ENVIRONMENTAL ETHIC

One form of the skeptic's objection goes like this. People, not trees or lakes, decide moral controversies. Therefore any decision regarding the fate of a tree or a lake will inescapably reflect our (human) preferences, not Nature's. Hence: all morality is irredeemably homocentric. This argument can be easily deflected. For although the premise is trivially true, the conclusion is severely misleading. Of course it is people who are going to make the moral decisions, not wolves. It is only the exercise of human power that is in question. And no ethicist denies that in the final analysis, people will do what they prefer doing (within their power), whether to other humans or to Nature. The point of morals is to get us to examine and, if need be, alter those preferences and restrain that power, to bring us around to preferring the morally right course of conduct over the morally wrong. The environmental ethicist recognizes that Nature's fate rests in human hands (that, after all, is why he or she is addressing humans). The goal is to persuade us, in making our choices, to provide a greater accounting for values other than straightforward human self-interest.

The second form the objection takes is tougher to answer. Suppose we agree that it is appropriate, in social decision-making, to make some allowance for Nature itself, as opposed to our competing interests *in* Nature. How do we proceed? What is there, transcending our own interests, to allow for?

To illustrate, suppose we are deciding whether to dam a wild river. No one doubts we would weigh the river's utility to us. We would begin by making calculations of the consumption and use values of the river in each alternative state. How do the revenues to be gained through electricity and irrigation compare to the loss of revenues through elimination of barge traffic? Next, we would add a "shadow price" to account for the fact that the revenues do not provide a complete picture, because many of the benefits of a river are so far-reaching and diffuse that not all those who benefit can be excluded from, and therefore forced to pay for, the value each derives. For

271

example, the shadow price of damming would include the lost beauty of the river's natural flow, measured by the amount people driving along its banks would pay to view it if we could cost-effectively charge them.

None of this is really controversial. What is in dispute is this: Is there anything we can say about the value of a natural object that transcends those calculations, specifically so that a social arrangement not warranted by our conventional homocentric measures—because we would be sacrificing "too much"—would yet be warranted by some morally corrected price?

The skeptic may acknowledge that there is more to morals than either utility (an action is "good" if it produces the greatest good for the greatest number of people) or conventionalism (an action is "good" if it commands the moral approval of the community). It was no justification for the Nazi war criminals to argue that their slave-labor policies conduced to the general welfare or that the majority of their countrymen approved of it. That would have been nonsense. But, the skeptic will say, we are prepared to introduce special "corrections"—to deviate from pure utility—in cases of that sort because the victims there were persons, and persons have properties universally recognized as morally significant: the capacities to feel pleasure and pain, to project a future, to understand what is happening to them, the freedom and power to exercise moral choice, a sense of justice, the fact of having been created, by biblical account, in God's image, and so forth. The thrust of these various factors, each in its own way, provides an avenue for treating persons as *morally considerable*, but leaves natural objects, it is said, in a position of moral nullities, mere resources for the benefit of persons.

This opens up an enormously complex theme: What does it take to be morally considerable? The environmentalist has to concede a wide range of difference between persons and "things." A person whose initial inclination is to perform some act can be persuaded to desist if he or she can be persuaded that it is the wrong thing to do. Such a dialogue is out of the question with natural objects—which is why a rampaging river or rhinoceros may do harm, but it cannot do *wrong*. All this can be expressed by conceding that, unlike humans, neither nonhuman nor natural objects are moral agents, holding preferences that are correctable by ethical reasoning. But it is a long leap from the fact that persons are the only audience of moral discourse—and therefore the only holders of moral obligation—to the conclusion that we alone are legitimate obligees. We are widely regarded as having moral obligations toward infants, the insane, the terminally comatose—none of whom is reciprocally obliged to us. Most of us agree that we have some duties to future generations (such as to avoid leaving them a lethally polluted planet). Yet the unborn are no better situated to "return the favor" than a contemporary river or tree. In sum, there is no reason to accept the premise that because a tree or mountain cannot be a moral obligor, it cannot, ipso facto, be morally considerate in the thinking of those—we humans—who *are* moral obligors.

The skeptic who concedes the obligor-obligee distinction may rejoin with another barrier: that to be considerable, a thing must be, at the least, a holder of moral rights. This is a position generally favored by writers and activists bent on establishing obligations toward animals, embryos, and fetuses, who typically maintain that their "clients" are holders of moral rights, even if they lack some of the properties requisite for full moral agency. What that may be—what is required for rights-holding—is a subject of considerable dispute. But whatever the disagreements, most of the literature appears to assume that, at the least, taking an interest in one's fate—some degree of self-consciousness—is a prerequisite for holding a moral right. A good deal might be said on behalf of chimpanzees on that score, but few will claim that mountains and rivers possess any self-awareness. Some environmentalists may respond by attacking the premise that self-awareness is a prerequisite of rights-holding.

I myself am ambivalent about the appropriateness of recognizing *moral rights* of nonpersons (as opposed to recognizing a set of *legal rights*, which I deem could be granted them quite intelligibly, if we care or feel morally compelled to do so). That is because moral rights seem most at home applied to moral agents who live in a common community—entities who are therefore capable of claiming, waiving, and trading their rights claims on their own volition. Such sophisticated choices are beyond the reach even of higher animals, and all the more so of trees. Hence, it seems more reasonable to speak of our duties toward an animal, rather than of the animal's rights.

Perhaps I capitulate too easily on the rights issue. (I have been told as much.) But one reason for my ready concession is that I am not sure rights matter nearly as much as the academic attention they receive suggests. Rights discourse is much in vogue today, but rights are not the sole basis for, nor even the keystone to, moral analysis. Consider for a moment the conventional good samaritan dilemma: a passerby spots an unconscious child drowning in a few inches of water. We need not reach the question whether the child in jeopardy has a *right* to the passerby's rescue, or even whether the passerby has a duty to the child, in order to say that it would be morally commendable for the passerby to rescue the child, that the passerby would, by saving the child, *be advancing some good.*

On a parity of reasoning, to support laws that give protection to a river or forest, one is not required to prove that the river or forest has a moral right to be so situated. There is prima facie support for the arrangement if one demonstrates that so to protect the environment is morally better than to destroy it. The environmental richness of the world, like the welfare of its *homo sapiens*, is one of the things to be counted in evaluating the goodness and evil of our actions.

This places our skeptic in a third, fall-back position. He may respond that even if we substitute some looser notion of "good" for one of rights as the key to moral evaluation, we have not detached ourselves from a thing's interests entirely. Any moral revision of our thinking still requires an accounting of the interests of the things on whose behalf our disposition is to be revised—otherwise, wherein lies the "good" in sacrificing our more obvious desires? If we discern some good in improving the lot of the poor, even at some sacrifice to the general welfare, it is because we can identify some interests that have been advanced. On the other hand, if we are asked to make a collective sacrifice in consideration of a river and are told to exclude from our thoughts any consideration of human welfare, where is the "good" in it to be found? Indeed, what are we being asked to think about?

This, it seems to me, identifies the proper heart of the controversy. But the objection, so refined, does not put the environmentalist out of court. In *Earth and Other Ethics* I outline at least five foundational bases on which an environmental ethic might be carried forward— that is to say, bases for argument that do not simply accept the utility preferences of the majority of contemporary human beings. The foundations range from one that accounts for future generations to one that rests on self-maximization, but which construes the "self" not as a conventional individual ego but as an extended self that incorporates a wide range of elements in one's environmental community. It can be maintained that some nonhuman things may be good intrinsically, that is, good not because of their use to us, but good in and of themselves.

Granted it is easier to think of something a moral observer might say about such "good" with respect to animals than on behalf of nonliving entities such as the Grand Canyon. In animals one can more readily identify the "goods" of familiar moral theory: a life that can be snuffed out, a plan that can be frustrated, a nerve that can transmit pain. The person who supports the moral considerableness of plant life and, even more so, of an inanimate object confronts the task of identifying some comparable basis, some "intrinsic worth" of something that cannot be killed, frustrated, or pained. That raises the obvious questions about "where do you draw the line?" ("What about my car battery, then?"). And a rejoinder seems hard to achieve without assuming a somewhat dogmatic stance, to find oneself declaring that "X's existence is good because . . . X's existence is good."

Yet all moral philosophies have their hardest going at the very same starting point, the identifying of the basic good or goods around which they will revolve. Utilitarianism adopts a particular psychological state, pleasure, as the intrinsic good—not, as we know, uncontroversially. Other philosophies adopt other foundational goods, for example, Kant's "good will." I do not find it insupportable to defend as an "intrinsic good" something without feeling, will, or interests (as Spinoza presumed God to be).

The Burdens an Environmental Ethic Faces

In fact, I do not find establishing a prima facie basis for moral considerableness to be the hardest part of the environmental ethicist's task. There are several foundational possibilities to work with, either separately or in combination. But if any of them is going to make much difference in how we think, it has several burdens to contend with.

First, what is being sought is not just a moral viewpoint that accounts for Nature in principle. We need a moral viewpoint rich enough to advance us through the ontological conundrums. By reference to what principles is the moral and legal world to be carved up into those "things" that count and those that don't? Will the unit of our concern be the individual ant, the anthill, the family, the phylum, the genus, or the ant's habitat? One cannot really avoid such quandaries by adopting a holistic or Gaian viewpoint, one that emphasizes the goodness of the planetary whole; that is, if the everything—the whole—is good, what can be bad? How, faced with any human choice, would a holist judge?

Second, suppose that we can do the carving up correctly, that is, identify those objects toward which some prima facie moral regard is justified, for example, perhaps a certain mountain. Even if moral obligations to a mountain are conceded to exist in principle, the question of how they can be discharged remains: How does one "do right by" a mountain?

Third, there are the distributional dilemmas. It is not enough to carve up the world, establishing what is to be of moral considerableness; nor is it enough to agree how that regard translates into prima facie good and bad acts. What are we to do in the case of conflicting indications? For example, suppose that, working with reference to one or another of the various nature-valuing viewpoints, a prima facie case can be made for preserving each of two species of animal—but we cannot preserve both. Do we favor the rare species of lower animal over the less exotic but "higher" one? It is easy to imagine a moral framework whose basic principle is "more life is better than less." One can imagine, too, support for the preservation of a singular

276

unsullied desert. What do we do in the face of an irrigation project that offers to transform the desert into a habitat teeming with life? The general problem is one of a moral judgment's strength. In other words, even if the continued existence of a species, or the state of a river, is granted to be "a good," how do we make adjustments for that moral fact in the face of conflict and of other, competing goods?

To illustrate with a present controversy, the most efficient way to catch tuna in the eastern Pacific is for the fleet to "set" its nets on dolphins. Not many years ago, before regulation, over 120,000 dolphins a year were entangling in the nets and dying, a number the Inter-American Tropical Tuna Convention (IATTC) has agreed to reduce below 20,000 in 1992 and is scheduled to phase downward (with observers being placed on all vessels) to 5,000 by 1999 (see chapter 6). But here is the moral catch. The dolphins and the ideally mature tuna travel together for the most part by themselves, outside the company of other creatures. Thus setting nets "on dolphins" takes big tuna, some dolphins, and little more. If human welfare (the satisfied appetite for tuna) holds steady and setting on dolphins is virtually eliminated, then to make up the deficit in tuna tonnage the fleets will have to move over onto schools of smaller immature yellowfin tuna, many of which have not yet reproduced. These smaller tuna, while not mixed with dolphins, swim in association with many different fish such as sharks and mackeral that when caught will be thrown overboard as mortally wounded "wastage." The result is that the measures required to hold tonnage constant and save dolphins, on the one hand, results in the death both of more tuna and more nontarget fish—perhaps with even more far-reaching impairment to the ocean ecosystem. (The species of dolphin involved are not endangered and even on the rebound at this point.) To be persuasive, therefore, the argument to reduce, even eliminate, dolphin kills (a move which enjoys considerable sympathy among environmentalists, including the author) has to be grounded on something like each dolphin (but not each tuna or nontarget fish) having a "right to life" (or, conversely, humankind having an absolute duty not to cause its death, even if that would involve, as it well might, giving up tuna salad).

While all of these questions are hard, they are no more formidable than those with which the proponents of any moral theory have to face. An environmental ethic, one that gives good, nonhomocentric guidance—dolphins versus tuna versus phyloplanktons—can be fleshed out. Aristotle's ethics (and Plato's and Confucius's and Kant's) have been honing and adapting over centuries. At this stage, the main impediment to the development of a nonhomocentric ethic involves uncertainties as to meta-ethics, that is, as to the constraints and ground rules by which a moral theory has to abide in general. Before those inclined to champion an environmental ethic make further headway, they will have to go back and reexamine some of the most fundamental issues of moral philosophy.

One Further Level of Abstraction: The Metaphysical Underpinnings

What I have in mind, specifically, is this. The orthodox assumption in ethics today is that the ethicist's task is to put forward and defend a single coherent body of principles, such as utilitarianism's greatest happiness for the greatest number or Kant's "categorical imperative," and to demonstrate how it—the one correct viewpoint—guides us through all moral dilemmas to the one right solution. This viewpoint, which I call "moral monism," implies that in defending, say, the preservation of a forest or the protection of a laboratory animal, we are expected to bring our argument under the same principles that determine our duties to kin or the punishment of terrorists. It suggests that moral considerableness is a matter of either-or; that is, it assumes there is a single salient moral property, such as sentience, intelligence, or having a conscious life, so that entities are (depending on whether they are blessed with the One Salient Property) *either* morally relevant (each in the same way, according to the same rules) *or* utterly inconsiderate, out in the moral cold. Mammals, because they display qualities *x*, *y*, and *z*, *count morally*; plants, because they lack them, don't.

But monism's ambitions, to unify all ethics within a single framework capable of yielding the One Right Answer to all our quandaries, collides with the variety of things whose considerableness commands some intuitive appeal: normal persons in a common moral community, persons remote in time and space, embryos and fetuses, nations and nightingales, beautiful things and sacred things. Some of these things we wish to account for because of their high degree of intelligence (higher animals); with others, sentience seems the key (lower life); on the other hand, the moral standing of membership groups such as nation-states, cultures, and species has to stand on some other footing, since the group itself (the species, as distinct from the individual whale) manifests no intelligence and experiences no pain. Other entities are genetically human, either capable of experiencing pain (advanced fetuses) or nonsentient (early embryos), but lack, at the time of our dealings with them, full human capacities.

Trying to force all these diverse entities into a single mold—the One Big, Sparsely Principled Comprehensive Theory—forces us to disregard some of our moral intuitions and contributes to dilating our primary principles into unhelpfully bland generalities. The commitment is not only quixotic; it imposes strictures on thought that stifle the emergence of more satisfactory approaches.

The alternative conception I have been propounding—what I call "moral pluralism"—invites us to conceive moral activities as partitioned into several distinct frameworks, each governed by distinct principles and logical texture. The framework appropriate for analyzing our obligations to the human unborn does not need to rest upon the same foundation as that which is appropriate to other species. Under such a pluralist approach, frameworks vary in regard to the fabric of their basic concepts. In some domains, as when we are considering actions that affect persons, we can speak in terms of rights and duties, of what is morally mandatory. In other domains, perhaps those encompassing butterflies and "lower" life, we may have to speak in less stern, more flexible judgments: not in terms of what is compulsory but what is morally welcome or permissible. In some domains we may have to abdicate the classic ambition to come up with one right answer and settle for a less binary "logic." In some areas, a moral framework may be satisfactory if it eliminates a subset

279

of "bad" choices, but is indeterminate over a broad range of other possibilities.

In other words, to accommodate and allow space for the growth of our intuitions about our relations with Nature, we have not only to choose among competing species, but even prior to that, among competing meta-ethical viewpoints. Do we embrace a moral framework in which all acts are good or bad insofar as they advance or diminish evaluated human welfare? Or are we going to buy into a version of the moral world in which *things* count morally—a version in which the question, "Is the earth the better with a billion more people or a billion more trees?" is a serious issue, not reducible to the average (or aggregate) happiness of hypothetical human populations?

That choice, the selection of an acceptable moral framework, cannot itself be *derived* from any prior moral postulates.* It is the answer to the question with which philosophy began—long before the contemporary preoccupation with evaluating acts and (less often) actors: *How ought one to live?* That and what sort of earth are we aiming for are the bedrock issues that lie at the intersection of philosophy and spirit. They are issues that are resolved less by formal choice, in which the mind is governed by appeals to consistency, than by lifestyle choices, in which we open ourselves to provocations of irony and humor, the experience of wilderness, and the echoes of the best traces of our indwelling terrestrial histories. And just a little humility would go a long way: ". . . *the gnat is older than he.*"

---

* To illustrate, once a geometer has stipulated her postulates, so as to produce a Euclidean or non-Euclidean geometry, proofs can proceed within its framework. But the choice of framework question, which geometry to select for which tasks—-for astronomy, architecture, etc.?—is not a question of geometry. To answer it, one must *step outside.*

✢ *Notes* ✢

Preface

1. The National Transportation Safety Board (NTSB) found that the captain's "judgment was impaired by alcohol during the critical period" of navigating the sound. He was not, however, convicted of piloting a vessel while intoxicated: the jurors rejected the state's claim because no blood alcohol test was given until eleven and a half hours after the accident. His only criminal responsibility was for misdemeanor pollution. See John H. Cushman, Jr., "Blame Is Placed for Valdez Spill," *New York Times*, August 1, 1990, A10. This charge was overturned by an appeals court on the grounds that he was immunized from prosecution under a special statute benefiting those who, like the captain of the *Valdez*, voluntarily reported spills to the Coast Guard. "Exxon Valdez Captain's Conviction is Overturned," *Los Angeles Times*, July 11, 1992, A18.

2. There was no shortage of potential culprits. The NTSB put part of the blame on the Coast Guard for having failed to monitor vigilantly and for having had inadequate ship traffic control and radar systems. John H. Cushman, Jr., "A Hard Look at Waterborne Traffic Control," *New York Times*, August 5, 1990, sec. 4, p. 4. Exxon paid a huge amount—perhaps $4.5 billion in fines, cleanup costs, criminal restitution, and damage settlements—but not until after it had filed a claim against the Coast Guard maintaining that the Coast Guard's negligent licensing of the pilot, its failure to warn the *Valdez* of imminent collision, its failure to monitor the ship, and its failure to advise mariners of radar limitations all contributed to the accident. "Coast Guard Negligent in Licensing Exxon Valdez Captain, Exxon Contends," *Platt's Oilgram News* 68 (October 4, 1990): 4.

3. See "A Lesson Learned, Again, at Valdez," *Science* 252 (1991): 371.

4. Ramanlal Soni, *Control of Marine Pollution in International Law* (Cape Town: Juta, 1985), x.

5. Edvard Hambro, "Some Legal Problems Concerning the Protection of the Human Environment," *Israel Law Review* 12 (1977): 1.

6. See Jared M. Diamond, "The Environmentalist Myth," *Nature* 324: 19–20, and chapter 10, infra.

7. For 1993 the Bush administration has proposed $1.3 billion for the U.S. Global Change Research Program (USGCRP), which is already far and away the world's largest program on global-change research. An additional

$30 million has been proposed for global-change-related economics research. For this side of the story, see Nancy G. Maynard, "Science: The Basis for Action on Global Change," *Arizona Journal of International and Comparative Law* 4 (1992): 35–46. Total U.S. funding for all environmental research has been put at $2.4 billion for 1992. "U.S. Funding for Environmental Research, 1992," *Science* 256 (1992): 1628.

8. Erick Eckholm, "A Casualty Review; AIDS, Fatally Steady in the U.S., Accelerates Worldwide," *New York Times*, June 28, 1992, E5.

9. "Treatable Maladies to Kill 100 Million Children," *Los Angeles Times*, December 13, 1989, A9. The story, which I have taped above my desk to help me keep perspective, goes on to report that 8,000 children die every day simply because they have not been immunized; nearly 7,000 die daily from dehydration and 6,000 from pneumonia. Ibid. See also UNICEF, *State of the World's Children* (New York: Oxford University Press, 1990). The World Health Organization reports that about one billion people—one out of every five persons on earth—are suffering from disease, poor health, or malnutrition. "The World, 1 Billion Ill or Hungry," *Los Angeles Times*, September 25, 1989, A2. It is uncertain exactly how accurate UNICEF's and WHO's figures are.

10. World Bank, *World Development Report 1992* (Oxford: Oxford University Press, 1992), 5, 49.

11. World Health Organization, "World Malaria Situation in 1989," *World Health Organization Weekly Epidemic Record*, 1991, no. 22: 157–63 and 1991, no. 23: 167–70; reprint, World Health Organization, June 1991, 4, suggests that while the million number is "the most frequently quoted" (which our research confirms) it is based on data from the 1950s, and has probably fallen from increased use of antimalarials, social development, and education.

12. See UNICEF, *State of the World's Children*, 18–19.

13. Ibid., 21. If UNICEF is correct, part of the tragedy is that with no additional expense, but only a shift in preventive and therapeutic techniques, these two diseases could be practically eliminated. Ibid.

14. Quoted in "Talk of the Town," *New Yorker*, June 29, 1992, 25.

15. 52 Fed. Reg. 47494 (1987). Those are the figures that were motivating the U.S. at the time of the Montreal and London negotiations. More recent data suggesting that the thinning has been more pronounced than anticipated have caused the EPA to revise its estimate, claiming that the thinning experienced thus far will cause 200,000 additional cases of fatal skin cancer in the next fifty years in the U.S. alone, even accounting for the

impact of the Montreal and London restrictions. See "Ozone Loss Hits Us Where We Live," *Science* 254 (1991): 645.

16. Worldwide, deaths might be in the range of twenty times larger than the 43,000 deaths the EPA estimated for the U.S. alone in the mid twenty-first century—say, 860,000 or fewer than 2,400 a day. Of course, there were other reasons to ban CFCs than to prevent cancer fatalities. For every fatal skin cancer case, there are roughly fifty nonfatal cases that are ordinarily readily treatable (my dermatologist removes mine in his office), but in the aggregate they constitute a considerable medical expense worldwide. The ultraviolet radiation whose increase is at the heart of the ozone fear is implicated in some other phenomena discussed in chapter 1. Also, CFCs are the most serious greenhouse gases, molecule for molecule, being 5,000 to 10,000 times more effective than carbon dioxide per unit of mass. Bert Bolin, "Man-Induced Global Change of Climate: The IPCC Findings and Continued Uncertainty Regarding Preventive Action," *Environmental Conservation* 18 (Winter 1991): 301. Still, for American industry alone the changeover costs from CFCs to their "best" substitutes, HCFCs, which themselves pose undetermined risks (discussed infra), amounts to $5 to $7 billion, and at a cost, in terms of equipment write-offs in the U.S. alone, of $19 to $34 billion, according to the Department of Energy; see "Choose Your Shade of Green," *The Economist*, April 14, 1990, 74. Even while the necessary investments are being made, more recent evidence on the unwanted side-effects of HCFCs already led, by 1992, to authoritative calls for phasing *them* out by 2005, long before the originally anticipated costs can be amortized. These costs would have gone a long way toward solving some of the dreadful, more immediate problems referred to in the text. My point is hardly to argue that the ozone agreements are wrong-headed. There just seems to be a peculiar moral, perhaps institutional blindness that demands examination.

17. See "Boom Time for Environmentalists," *WorldWatch*, November/December 1988 (attributing the largest ever surge in environmental-group membership to fears of greenhouse warming and the *Valdez* disaster).

18. See, for 1991 figures, Robert Pear, "U.S. Ranked No. 1 in Weapons Sales," *New York Times*, August 11, 1991, A8; for 1992 figures, idem, "U.S. Sales of Arms to the Third World Declined by 22% Last Year," *New York Times*, July 21, 1992, A16.

19. See Peter H. Stone, "Defense Exporters' Secret Weapons," *Legal Times*, February 25, 1991, 1. More recently the industry has even pressed for earmarking some of the supposed defense conversion funds to subsidize foreign sales.

20. In fact, Germany, in recently announcing constraints on foreign arms sales, specifically alluded to distortions of LDC agendas that the sales were reinforcing. And in 1992 both the International Monetary Fund and World Bank began sending signals that in apportioning developmental assistance they will (finally) account for an aid applicant's military budget. In regard to the policies of the developed countries, some background has to be borne in mind. In the case of the U.S., most of its foreign arms sales went to the Saudis—a special case. And of course the special treatment the defense industry receives in capitals across the world cannot be approached as though success were entirely the product of effective power lobbyists and the specter of unemployment; all nations have an understandable interest in nursing their domestic defense industry through peace's hard times. On the other hand, my impression is that the budding environmental cleanup industry, like most, employs more people, per dollar sales, than defense.

CHAPTER I
DIAGNOSIS

1. Alan Gregg, "A Medical Aspect of the Population Problem," *Science* 121 (1955): 682–83.
2. See J. E. Lovelock, *Gaia: A New Look at Life on Earth* (Oxford: Oxford University Press, 1979).
3. See Nathan Keyfitz, "The Growing Human Population," *Scientific American* 261 (September 1989): 119–26.
4. "Making Polluters Pay," *The Economist*, September 2, 1989, Survey, 7.
5. See Peter Waldman, "Water Often Is a Divisive Issue in the Fractious Middle East," *Wall Street Journal*, October 30, 1991, A10; "Water Resources: Regional Water Crisis Could Trigger Next Middle East War, Observers Say," *International Environment Reporter (BNA)* 13 (1990): 413–14.
6. The World Bank, *World Development Report 1992*, 5.
7. Ibid., 48–49.
8. World Resources Institute et al., *World Resources 1988–89* (New York: Basic Books, 1988), 111.
9. Norman Myers, author of *The Primary Source: Tropical Forests and Our Future* (New York: Norton, 1992), quoted by A. Kent MacDougall, "Solemn Transition; Worldwide Costs Mount as Trees Fall," *Los Angeles Times*, June 14, 1987, I1, I18.
10. See Paul E. Hagen, "The International Community Confronts Plastics Pollution from Ships: Marpol Annex V and the Problem That Won't Go

Away," *American University Journal of International Law and Policy* 5 (1990): 425–96.

11. Eugene Linden, "Putting the Heat on Japan," *Time*, July 10, 1989, 52, quoting a biologist from Earthtrust.

12. Figures released at the 1989 Annual Meeting of the International Fisheries Conference estimated the yearly death of 80,000 marine mammals and a million seabirds. Timothy Egan, "New Evidence of Ecological Damage Brings a Call to Ban Drift-Net Fishing," *New York Times*, September 6, 1989, A24.

13. Marlise Simon, "Fish Nets Trap Dolphins in the Mediterranean, Too," *New York Times*, September 6, 1989, A1.

14. *See* International Union for the Conservation of Nature (IUCN), *World Conservation Strategy* (IUCN, United Nations Environmental Programme and World Wildlife Fund, 1980), §4.

15. See "What Else We Found," *Consumer Reports* 57 (February 1992): 103.

16. See Stephanie Simon, "Fears over Nazi Weapons Leaking at Bottom of Baltic," *Los Angeles Times*, July 18, 1992, A3. An additional 200,000 tons of weapons were dumped in other locations, principally the English Channel and off the coasts of Denmark, Sweden, and Norway.

17. See "Atomic Waste Reported Leaking in Ocean Sanctuary off California," *New York Times*, May 7, 1990, B12 (about one-fourth of 47,500 fifty-five-gallon drums dumped between 1947 and 1970 off San Francisco had ruptured, threatening to contaminate local fish resources). How much alarm the potential leakage warrants is controversial. See F.G.T. Holliday, "The Dumping of Radioactive Waste in the Deep Ocean: Scientific Advice and Ideological Persuasion," in *The Environment in Question*, ed. David E. Cooper and Joy A. Palmer (London and New York: Routledge, 1992), 51–64, 56–59. In all events, contracting parties to the London Dumping Convention agreed to a moratorium on marine disposal of radioactive wastes in the 1980s; the moratorium is due for revision in 1993. See "Opponents to Nuclear Waste Dumping Ban Want Global Action, Senior IMO Official Says," *International Environment Reporter (BNA)* 15 (1992): 353.

18. Patrick E. Tyler, "Soviets' Secret Nuclear Dumping Causes Worry for Arctic Waters," *Los Angeles Times*, May 4, 1992, A1. Was the U.S. Navy engaged in the same practices, if on a smaller scale? A retired Navy pilot claims (and the Navy denies) that he was ordered to toss four to eight beer-barrel-sized canisters into the Atlantic in 1947. The U.S. formally suspended offshore radioactive waste dumping in 1970. Donald P. Basker, "Congress

to Probe Charges of Nuclear Dumping in Ocean," *Washington Post*, January 4, 1981, A11. Removing the canisters, should that be deemed otherwise prudent, poses risks of rupturing them in the process.

19. See Gina Kolata, "Tree Yields a Cancer Treatment, but Ecological Costs May Be High," *New York Times*, May 13, 1991, A1; Lawrence K. Altman, "Antibiotics Natural to Frogs Battle Some Disease Agents," *New York Times*, July 31, 1987, A1.

20. See Norman Kempster, "Plant Safety Gains Seen as Too Slow; Eastern Europe: Western Experts Plod Ahead but Concede That Danger Remains Severe," *Los Angeles Times*, March 26, 1992, A4 (reporting that over two dozen reactors in the region are operating without containment vessels and other safety devices required in the West). See also "Other Chernobyls Waiting to Happen? The Fear Is of an Archipelago of Nuclear Volcanoes," *Los Angeles Times*, March 25, 1992, B6 (Greenpeace estimating the risks of a meltdown at 27 percent over a five-year period).

21. See David P. Hackett, "An Assessment of the Basel Convention on the Control of Transboundary Movements of Hazardous Waste and Their Disposal," *American University Journal of International Law and Policy* 5 (1990): 294.

22. See *Superfund II: A New Mandate*, BNA Special Report, February 13, 1987, 8.

23. See Hackett, "Assessment of the Basel Convention." This traffic is now being made the subject of the Basel Convention on the Control of Transboundary Movements of Wastes and Their Disposal. See chapter 3.

24. See "Ozone Destruction Worsens," *Science* 252 (1991): 204 (reporting on new satellite data that suggest that the thinning over the U.S. is double the previously reported rate); "Ozone Loss Hits Us Where We Live," *Science* 254 (1991): 645 (reporting evidence of a 3 percent thinning in north temperate latitudes in the past decade). Interestingly, UNEP, which has no reason to downplay the risks of ozone thinning, has estimated that a 3 percent decrease in the ozone shield increases the risk of skin cancer about as much as if someone moved two or three degrees latitude closer to the equator, about the distance from Boston to Philadelphia. United Nations Environmental Program, *The Greenhouse Gases* (Nairobi: UNEP/Gems Environment Library, 1987), 24.

25. The EPA has raised the possibility that UV-B exposure may suppress the immune system's response to certain diseases, like herpes simplex and leishmaniasis, a skin disorder. However, the EPA also conceded there was

no mechanism to explain the immune response, that it was impossible to quantify this effect (*Federal Register* 52 [1987]: 47495), and did not cite any authorities in the *Federal Register* to back up this speculation. Some researchers have proposed mechanisms by which UV-B radiation might suppress immune responses. See R. A. Daynes et al., "Immunomodulation by Ultraviolet Radiation: Prostaglandins Appear to Be Involved in the Molecular Mechanisms Responsible for UVR-Induced Changes in Immune Function," in *Effects of Changes in Stratospheric Ozone and Global Climate*, ed. James G. Titus, 2:63 (USEPA: Washington, D.C., 1986); Craig A. Elmets et al., "Stratospheric Ozone Depletion: Immunologic Effects on Monocyte Accessory Function in Humans," in ibid, 87. Then, too, there is concern that UV-B might damage plant productivity. Soybeans exposed to conditions simulating 16 to 25 percent ozone depletion lost 25 percent of expected productivity. Two thirds of two hundred other tested plants were somewhat UV-sensitive, though they were only tested in the laboratory, not the field. Ibid. For further discussion, see Alan H. Teramura, "Overview of Our Current State of Knowledge of UV Effects on Plants," in *Effects of Changes*, 1:165.

26. See R. C. Smith et al., "Ozone Depletion: Ultraviolet Radiation and Phytoplankton Biology in Antarctic Waters," *Science* 255 (1992): 952 (estimating a minimum 6 to 12 percent decrease in primary phytoplankton production that could be associated with ozone depletion).

27. Reportedly, a fifteen-day exposure to UV-B levels 20 percent higher than normal can kill off anchovy larvae to a depth of 10 meters. World Resources Institute et al., *World Resources 1988–89*, 175, n. 122, citing B. Thompson, "Is the Impact of UV-B Radiation on Marine Zooplankton of Any Significance?" in *Effects of Changes*, 2: 203–9, and table I, 205.

28. Kevin Fox and Tom Lutgen, "Today on the Planet," *Los Angeles Times*, May 26, 1992, H6.

29. See "A Greener Bank," *The Economist*, May 23, 1992, 79–80.

30. "Tracing Arctic Haze," *New York Times*, January 15, 1991, C6. (During the winter and spring Arctic pollution levels can at times rival the pollution levels of large cities.)

31. Intergovernmental Panel on Climate Change, *Policymakers' Summary of the Potential Impacts of Climate Change: Report Prepared for IPCC by Working Group II*, June 1990, 1 (hereafter IPCC Working Group II) (doubling will occur between 2030 and 2050).

32. By a Swedish chemist, Svante Arrhenius, who not only discerned the dynamics, but even then foresaw the possibility that the burning of fossil

fuels would warm the planet. See Wallace S. Broecker, "Global Warming on Trial," *Natural History*, April 1992, 6.

33. However, a more recent revised IPCC report predicts that the global temperature increase will be 20 to 30 percent lower than the IPCC projections in 1990. See *International Environment Reporter (BNA)* 15: 350–51.

34. See William K. Stevens, "Humanity Confronts Its Handiwork: An Altered Planet," *New York Times*, May 5, 1992, C1.

35. Intergovernmental Panel on Climactic Change, *Policymakers' Summary of the Scientific Assessment of Climate Change: Report Prepared for IPCC by Working Group I*, June 1990, 1 (hereafter IPCC Working Group I). IPCC Working Group II reported slightly differently, emphasizing a doubling of GHG concentration by 2030 to 2050, with a "consequent" (time frame unspecified) increase of global mean temperature of $1.5°$ to $4°–5°$ C. IPCC Working Group II, 1, 8. Group II left it unclear whether they expected that increase to occur *contemporaneously* with the doubling, or *eventually* from the doubling.

36. Seidel and Keyes, *Can We Delay a Greenhouse Warning?*, chap. 1, p. 5.

37. Of course, we also have today the benefit of technology that enables us to respond to changes more quickly than we could in the past.

38. IPCC Working Group II, 1.

39. G. Dimock, "The Reinsurer's View," in *The Greenhouse Effect: Implications for Insurers* (London: Insurance and Reinsurance Research Group, 1989), 117.

40. See Andrew F. Dlugolecki, "Natural Catastrophes Arising from the Greenhouse Effect," *Journal of Insurance Institutions of London* 78 (1990): 49–57 (placing frost damage first in a list of greenhouse perils).

41. The National Academy of Sciences Adaptation Panel report indicates that we should expect *new* pests rather than *more*; some pests flourish in the heat, others in cold. Committee on Science, Engineering, and Public Policy, *Policy Implications of Greenhouse Warming* (Washington, D.C.: National Academy Press, 1992), 560–61. The entire biodiversity issue is also controversial; compare Charles C. Mann, "Extinction: Are Ecologists Crying Wolf," *Science* 253 (1991): 736–38, with Robert L. Peters and J. P. Myers, "Preserving Biodiversity in a Changing Climate," *Issues in Science and Technology* 7 (Winter 1991–92): 66–72.

42. See Marshall Fisher and David E. Fisher, "The Attack of the Killer Mosquitoes," *Los Angeles Times Magazine*, September 15, 1991, 30–35. The influence of viruses and pests on human history is well documented by William H. McNeill, *Plagues and Peoples* (New York: Anchor Books, 1976).

43. J. S. Hoffman et al., "Future Global Warming and Sea Level Rise," in *Iceland Coastal and River Symposium*, ed. G. Sigbjarnarson (Reykjavik, Iceland: National Energy Authority, 1986), cited in James G. Titus, "The Causes and Effects of Sea Level Rise," in *Effects of Changes*, 1: 223. Roger Revelle has estimated that a 4.2° C warming by 2050–2060 would raise sea level 30 centimeters; Revelle, "Probable Future Changes in Sea Level Resulting from Increased Atmospheric Carbon Dioxide," in *Changing Climate* (Washington, D.C.: National Academy Press, 1983), cited in Titus, "Causes and Effects," in *Effects of Changes*, 1: 222. The ill-understood lag time between air warming and ocean warming makes all these estimates highly uncertain.

44. James G. Titus, "Causes and Effects," in *Effects of Changes*, 1: 221, table 1, citing J. T. Hollin and R. G. Barry, "Empirical and Theoretical Evidence Concerning the Response of the Earth's Ice and Snow Cover to a Global Temperature Increase," *Environment International* 2 (1979): 437–44.

45. Working Group I's estimates, 6 centimeters per decade (with an uncertainty range of 3 to 10 centimeters per decade) with 65 centimeters expected by 2100, were slightly lower than, but not wildly out of line with, Group II, which came up with the estimate of one meter by 2100. IPCC Working Group I, i; IPCC Working Group II, 1. A previous rise of 30 centimeters by 2035 had been estimated on the effect of a 1° to 2° C temperature rise. "The Warming Globe," *The Economist*, September 2, 1989, Survey, 12. Hoffman et al., examining a variety of possible scenarios of future emissions of greenhouse gases and global warming, estimate that a warming of between 1° and 2.6° C could result in a thermal expansion contribution to sea level of between 12 and 26 centimeters by 2050. They also estimated that a global warming of 2.3° to 7.0° C by 2100 would result in thermal expansion of 28 to 83 centimeters by that year. Hoffman et al., "Future Global Warming," cited in Titus, "Sea Level Rise," in *Effects of Changes*, 1: 223.

46. In the United States, the Carolina coast is most frequently cited as being at risk. In Egypt, the area that stands to be inundated in a one-meter rise is occupied by approximately 16 percent of the nation's estimated 49 million population. James Broadus et al., "Rising Sea Level and Damming of Rivers: Possible Effects in Egypt and Bangladesh," in *Effects of Changes*, 4: 172. In Bangladesh, approximately 9 percent of the nation's present 93 million people would be directly affected by a one-meter rise, and 27 percent of the total population would be affected by a 3-meter rise. There, exposure of the population to storm surge would be an even more grave concern. On average, 1.5 severe cyclonic storms already assault the country each year

with surges invading as far as 160 kilometers inland, exacting large tolls in human life. Ibid., 175.

47. On the likelihood that increased sea surface temperatures (SST) will increase storm generation, see ibid., 154, and authorities cited. The El Niño/Southern Oscillation of 1982–83 in the eastern Pacific reportedly resulted in a doubling of the average frequency of hurricanes, from ten to twenty. Low-lying coastal regions of the world, which are presently subjected to $6 to $7 billion in damage each year, as well as 20,000 deaths worldwide, are at risk of increased frequency and intensity of tropical cyclones. Ibid. On the other hand, the more recent IPCC findings question whether the 26° C ocean surface temperature critical for storm generation today would remain critical when the ambient temperature is elevated. It concludes, "Although the theoretical maximum intensity is expected to increase with temperature, climate models give no consistent indication whether tropical storms will increase or decrease in frequency or intensity." IPCC Working Group I, 18.

48. See "In Arctic, a Toxic Surprise," *Los Angeles Times*, June 18, 1991, A1.

49. Stephen Schneider, "The Changing Climate," *Scientific American* 261 (September 1989): 73.

50. Richard P. Turco et al., "Climate and Smoke: An Appraisal of Nuclear Winter," *Science* 247 (1990): 166.

51. See, generally, Arthur Westing, *Environmental Warfare* (London and Philadelphia: Taylor and Francis, 1984).

52. See Larry B. Stammer, "Geopolitical Effects of Global Heating Gauged," *Los Angeles Times*, February 10, 1992, A1.

53. See "Water Resources" (note 5).

54. See Waldman, "Water Often Is a Divisive Issue" (note 5).

55. Ibid.

56. Quoted in H. J. Taubenfeld and R. F. Taubenfeld, "Modification of the Human Environment," in *The Future of the International Order*, ed. Cyril E. Black and Richard Falk (Princeton: Princeton University Press, 1972), 4: 128.

57. The best comprehensive exposition of the positive view remains Julian L. Simon and Herman Kahn, *The Resourceful Earth* (Oxford and New York: Basil Blackwell, 1984).

58. Globally, however, the trend remains upward. See World Resources Institute et al., *World Resources 1988–89*, 164.

59. See C. Flavin, "Carbon Rate Takes a Breather," *WorldWatch* 5 (May–June 1992): 33 (citing a decline in rates of increase, from 4.6 percent annu-

ally in the 1950s and 1960s to 2.5 percent in the 1970s and 1.2 percent in the 1980s, some of which, most recently, may be attributed to decreased economic activity). The recent decline in rate of growth of methane emission is reported in *International Environment Reporter (BNA)* 15 (1992): 493–94.

60. "A Day in the Life of Mother Earth; The Good News . . .," *Los Angeles Times*, May 26, 1992, H4–5.

61. Donald G. McNeil, Jr., "How Most of the Public Forests Are Sold to Loggers at a Loss," *New York Times*, November 3, 1991, sec. 4, p. 2.

62. Paul Ehrlich and Anne Ehrlich, *The End of Affluence* (Rivercity, Mass.: Rivercity Press, 1974), 21.

63. Simon and Kahn, *The Resourceful Earth*, 2. But see note 64.

64. The World Resources Institute estimates that from the mid-1960s to the mid-1980s, world food production increased at an annual rate of 2.4 percent, with grain production increasing at 2.9 percent; these figures must be put in the context of world population growth, which has been slowing to less than 2 percent. See Pierre R. Crosson and Norman J. Rosenberg, "Strategies for Agriculture," *Scientific American* 261 (September 1989): 132. However, figures for 1989–91 are not as favorable. Global consumption exceeded global production in 1987, 1988, and 1989. Global cereal stocks dropped 100 million metric tons in 1988, and are projected to have dropped 22 million tons in 1989. Two causes of these statistics were a drought in the U.S. and reforms of U.S. agricultural policies designed to slow production. World Resources Institute, *World Resources 1990–91*, 86. WRI suggests elsewhere that the rate of increase of productivity will begin to decline. First, the best lands for agriculture have been developed. Second, the greatest developments in agriculture, such as fertilizers, pesticides, machinery, and use of high-yield crops, have already produced the biggest gains of the Green Revolution. Ibid., 84.

65. Ehrlich and Ehrlich, *The End of Affluence*, 28–29. This gloomy prediction follows on the heels of meteorological data suggesting (wrongly) that the monsoons will not return to India in this century, from which we were to prepare for "hundreds of millions of additional people perishing." Ibid., 28.

66. Stephen Schneider, *The Genesis Strategy* (New York: Plenum Press, 1976), 90 (relating, in the course of expressing general concern about climate variability, warnings about global cooling).

67. See note 33.

68. Warren Brookes, "Coolheadedness on Warming," *Washington Times*, April 17, 1991, G1.

69. Ronald A. Taylor, "NASA Satellites Find No Sign of 'Greenhouse' Warming," *Washington Times*, March 30, 1990, A1.

70. Warren Brookes, "Man and Earth: Adaptation? Exploitation? Warnings More Hype Than Measurable Fact," *Washington Times*, April 20, 1990, H7.

71. See H. E. Landsberg, "Global Climate Trends," in Simon and Kahn, *The Resourceful Earth*, 274.

72. See Joyce E. Penner, "Effects of Aerosol from Biomass Burning on the Global Radiation Budget," *Science* 256 (1992): 1432–34.

73. See "Air Pollution: Stratospheric Ozone Hole Found to Occur in Mid-Hemisphere," *International Environment Reporter (BNA)* 14 (1991): 590.

74. See Patrick J. Michaels and David E. Stooksbury, "The Failure of the Popular Vision of Global Warming," *Arizona Journal of International and Comparative Law* 9 (1992): 65.

75. Ibid., 61.

76. Ibid., 65.

77. A. Yanshin, "Is the Greenhouse Effect Really That Dangerous?" *Scientific World* 34, no. 3 (1990): 15.

78. See UNEP, *The Greenhouse Gases* (Nairobi: UNEP, 1987), 32. See also "Bringing Down the Sea Level Rise," *Science* 246 (1989): 1563.

79. See UNEP, *The Greenhouse Gases*, 22, reporting that, on the basis of an approximately 4° C temperature rise, "three of the most recent model predictions suggest that overall precipitation will increase by between 7 and 11 percent."

80. Quoted in A. Berrie Pittock, "The Carbon Dioxide Debate: Reports from SCOPE and DOE," *Environment* 29 (January–February 1987): 29.

81. Andrew Bakun, "Global Climate Change and Intensification of Coastal Ocean Up Welling," *Science* 247 (1990): 198.

82. See Mark Perlman, "The Role of Population Projections for the Year 2000," in Simon and Kahn, *The Resourceful Earth*, 52–66.

83. See Lovelock, *Gaia*.

84. It seems incredible, but our understanding of the carbon cycle remains so spotty that even today we simply cannot account for the 7 billion tons of carbon dioxide that we spew into the atmosphere annually. Reportedly, 3.5 billion tons appear to remain there. Two billion tons may be storing in the ocean. That leaves one or more billion tons at large, even after fifteen years of research. See R. A. Kerr, "Fugitive Carbon Dioxide: It's Not Hiding in the Ocean," *Science* 256 (1992): 35. In fact, there may be a ten-

dency to be less alarming in the more recent reports on greenhouse warming than in the original reports. See, for example, Michael Schlesinger and Xingjian Jiang, "Revised Projection of Future Greenhouse Warming," *Nature* 350 (1991): 219–21.

85. However, the most recent report on the prospects of reducing atmospheric CO_2 through artificial stimulation of phytoplankton growth is pessimistic. "Seeding with Iron," *New York Times*, January 22, 1991, C5 (success would be constrained by limited capacity of water to store CO_2).

86. See, generally, William J. Broad, "Scientists Dream of Bold Remedies for Ailing Atmosphere," *New York Times*, August 16, 1988, A19.

87. Schneider, "The Changing Climate," 73.

88. See "Fugitive Carbon Dioxide" (note 84).

89. William H. Calvin, "Greenhouse and Ice House," *Whole Earth Review* (Winter 1991), 106–11. Michaels and Stooksbury advance a case on which net cooling might reasonably be expected—although on a noncatastrophic scale. Michaels and Stooksbury, "Failure of the Popular Vision," 62–63.

90. See Lawrence K. Altman, M.D., "Fearful of Outbreaks, Doctors Pay New Heed to Emergency Viruses," *New York Times*, May 9, 1989, C3 (reporting that $150 million annually would underwrite a worldwide system to provide early detection).

91. See J. H. Gibbons, P. D. Blair, and H. L. Gavin, "Strategies for Energy Use," *Scientific American* 291 (September 1989): 141.

92. Cynthia Pollock Shea, "Protecting the Ozone Layer," in Lester R. Brown et al., *State of the World 1989*, ed. Linda Starke (New York and London: W. W. Norton, 1989), 85.

93. Although if strict emissions laws were passed, an inventor could look toward industries that required the patented device, rather than to the public till, as the source of reward.

94. Whether technological optimism is warranted is another question; see James E. Krier and Clayton Gillette, "The Uneasy Case for Technological Optimism," *Michigan Law Review* 84 (1985): 405–29, pointing out how imperfections in politics may impede the development of technological responses.

95. Thomas E. Graedel and Paul J. Crutzen, "The Changing Atmosphere," *Scientific American* 261 (September 1989): 63. To the same effect, see Wallace S. Broecker, "Unpleasant Surprises in the Greenhouse?" *Nature* 328 (July 9, 1987): 123–26 ("We play Russian roulette with climate [and] no one knows what lies in the active chamber").

CHAPTER II
THE CONDITION OF THE EARTH

1. See chapter 10.
2. See chapter 4.
3. See chapter 4.
4. See chapter 3.
5. See Richard A. Falk, "Environmental Warfare and Ecocide," *Bulletin of Peace Proposals* 4 (1973): 1–17.
6. Roderick Nash, *Wilderness and the American Mind* (New Haven, Conn.: Yale University Press, 1982), 344.
7. Quoted in Charles M. Hassett, "Air Pollution: Possible International Legal and Organizational Responses," *New York University Journal of International Law and Politics* 5 (1972): 38.
8. One of the few correct write-ups of this complex device, and its history, is Marilyn Post, "The Debt for Nature Swap: A Long-Term Investment in the Economic Stability of Less Developed Countries," *International Lawyer* 24 (1990): 1071–98.
9. See "U.S. Trade Mission Pushes Debt-Nature Swaps," *International Environment Reporter (BNA)* 14 (1991): 377.
10. See "Poland: Government Tries to Persuade nations to Swap Debt for Environmental Protection," *International Environment Reporter (BNA)* 14 (1991): 327.
11. Larry B. Stammer, "Saving the Earth: Who Sacrifices? Environment Could Lose Out in Face of Economic Realities," *Los Angeles Times*, March 13, 1989, sec. 1, pp. 1, 16.
12. See Amory B. Lovins and Ashok Gadgel, *The Megawatt Revolution: Electric Efficiency and Asian Development* (Snowmass, Colo.: Rocky Mountain Institute, 1991).
13. Brazil's resettlement programs required settlers to clear-cut tropical forest to establish ownership, even though the land so cleared only supported crops for a few years. Sandra Postel, "Halting Land Degradation," in Brown et al., *State of the World 1989*, 29. Brazil's deforestation rate dropped from 11,580 square miles per year in 1985 to 4,299 square miles per year by August 1991, a 63 percent decrease. James Brooke, "The Road to Rio—Setting an Agenda for the Earth; Homesteaders Gnaw at Brazil Rain Forest," *New York Times*, May 22, 1992, A1. Brazilian and American scientists explain the welcome decline by pointing to the suspension of tax

incentives for ranching and logging in the Amazon, a public information campaign, fines against cutters, and heavy rain, which has reduced forest fires. Ibid.

14. Marc Reisner, "The Next Water War: Cities Versus Agriculture," *Issues in Science and Technology* 5 (Winter 1988–89): 98. For a more extensive, and no less colorful, exposition of the misdevelopment of water law in the American West, see Reisner, *Cadillac Desert* (New York: Penguin Books, 1988).

15. See Postel, "Halting Land Degradation," 34–40, and table 2.4, emphasizing the astonishing gains available through institutional reforms. A similar point is made in the *Scientific American* special issue on "Managing Planet Earth" (September 1989), that the main task in agriculture is developing institutional mechanisms to make individual farmers correctly value the resources they are employing. Crosson and Rosenberg, "Strategies for Agriculture," 128–35.

16. Philip Shabecoff, "Ecologists Press Lending Groups," *New York Times*, October 29, 1986, A1.

17. In theory, countries also have a wide range of diplomatic recourse. A nation offended by another nation's conduct can always oust the offender's diplomats, close its air space, and so on. We have never been willing to take such moves in response to human rights violations. We are unlikely to do so in response to a nation's internal environmental policy. See, in general, Lisa Martin, *Coercing Cooperation* (Princeton: Princeton University Press, 1992).

18. "Japan Agrees to End Endangered Hawksbill Turtle Imports After '92," *Los Angeles Times*, June 19, 1991, A15.

19. H.R. 2782, *Congressional Record*, 101st Cong., 1st sess., 1989, 135: H3317. The bill has not been introduced in the 102d Congress. More recently, Senator David L. Boren of Oklahoma has introduced a bill, S. 984, that would require the imposition of special duties on imports from any nation that does not "impose and enforce effective pollution controls and environmental safeguards."

20. GATT panel report, "United States-Restrictions on Imports of Tuna," no. DS21/R, reprinted in *International Legal Materials* 30 (1991): 1594–1623.

21. World Bank, *World Development Report 1992*.

22. "A Greener Bank," *The Economist* 324 (May 23, 1992): 79.

Chapter III
Transboundary Pollution

1. The statement in text reflects the general posture of the law. In the more sophisticated view of the economist, the placement of "blame" between *A* and *B* is neither so clear-cut nor as critical an issue. See the discussion of the Coase theorem in this context in chapter 7.

2. The language is that of §601 of the Restatement (Third) of the Foreign Relations Law of the United States. Compare the wording of Stockholm Principle 21. Note the qualification, "to the extent practicable under the circumstances."

3. See David A. Wirth, "The World Bank and the Environment," *Environment* 28 (December 1986): 33.

4. The ECE is one of three major overlapping (and often confusing) groupings of major industrialized nations. One, the European Community (EC), appears to be forming out of Belgium, France, Germany, Greece, Ireland, Italy, Luxembourg, the Netherlands, Portugal, Spain, and the United Kingdom. Denmark, slated to enter the EC, appears to be rejecting entrance. The Organization for Economic Co-operation and Development (OECD), headquartered in Paris, includes all the EC countries, and, in addition, Australia, Austria, Denmark, Finland, Iceland, Japan, New Zealand, Norway, Sweden, Switzerland, Turkey, the United States, and Yugoslavia (the last, now tearing apart, named as an associate member). The U.N. Economic Commission for Europe (ECE) has some of the OECD countries, plus an Eastern bloc representation; it consists of Albania, Austria, Belarus, Belgium, Bulgaria, Canada, Cyprus, Czech and Slovak Federal Republic, Denmark, Estonia, Finland, France, Germany, Greece, Hungary, Iceland, Ireland, Israel, Italy, Latvia, Liechtenstein, Lithuania, Luxembourg, Malta, Maldova, the Netherlands, Norway, Poland, Portugal, Romania, the Russian Federation, San Marino, Spain, Sweden, Switzerland, Turkey, Ukraine, United Kingdom, United States, and Yugoslavia.

5. See "Transboundary Pollution: Treaty Signed by Twenty-Six Nations Sets Way to Protest Cross-Border Pollution," *International Environment Reporter (BNA)* 14 (1991): 99.

6. Natural Resources Defense Council v. Nuclear Regulatory Commission, 647 F.2d 1345 (1981).

7. National Organization for the Defense of Marijuana Law (NORMAL) v. Dept. of State, 452 F. Supp. 1226 (1978). In another case arising out of U.S. participation in construction of the Darien Gap Highway in Panama and

Colombia, the agencies did prepare an EIS, but environmentalists challenged as inadequate its examination of impacts on hoof-and-mouth disease, alternative routes, and the adverse effects on local Indian tribes. The Court of Appeals ruled that all three matters had been adequately covered. Sierra Club v. Adams, 8 Envtl. L. Rep. 1011 (D.C. Cir. 1978).

8. Executive Order 12114 (January 1979).

9. S. 1278, 102d Cong., 1st sess. (Lautenberg (D) New Jersey).

10. This has been one reading of the Lac Lanoux arbitration, instituted by Spain (without success) over changes that France made in the lake; Spain asserted that France was obliged at least to give it prior notice of its plans. See Allen L. Springer, *The International Law of Pollution: Protecting the Global Environment in a World of Sovereign States* (London and Westport, Conn.: Quorum Books, 1983), 146. Springer's treatment of the options reviewed in the text is highly recommended for those seeking a more detailed accounting.

11. See note 4.

12. OECD, *Strengthening International Cooperation on Environmental Protection in Transfrontier Regions*, Recommendations C(78)77 (Final), adopted September 21, 1978. The same idea has been proposed for nations intending to locate a nuclear power plant in a border-adjacent site. Gunther Handl, "An International Legal Perspective on the Conduct of Abnormally Dangerous Activities in Frontier Areas: The Case of Nuclear Power Siting," *Ecology Law Quarterly* 7 (1978): 30–39.

13. Paul Lewis, "Atomic Power Safety Steps Approved," *New York Times*, September 27, 1986, sec. 1, p. 36. See the 1986 Convention on Early Notification of a Nuclear Accident, in *International Environment Reporter (BNA)* Reference File 21: 3401.

14. World Charter for Nature Draft Principle 20. The draft had simply provided: "Relevant information must be supplied by States on activities or developments within their jurisdiction under their control whenever . . . they have reason to believe that such information is needed to avoid the risk of significant adverse effects on the environment in areas beyond their national jurisdiction." U.N. Doc. A/Conf. 48/P.C./17, para. 78 (1972).

15. L.F.E. Goldie, "International Maritime Environmental Law Today— An Appraisal," in *Who Protects the Ocean? Environment and the Development of the Law of the Sea*, ed. J. L. Hargrove (St. Paul, Minn.: West, 1975), 102–6, and Jan Schneider, in *World Public Order of the Environment: Towards an International Ecological Law and Association* (Toronto and Buffalo: University of Toronto Press, 1979), 159, give some background on how different na-

tions' varying interpretations of the simple and worthy language of Draft Proposal 20 led to its being watered down.

16. Proponents of such a rule illustrate it with the claim that while a lower riparian (a "downriver" state) has no clear legal right to enjoin an upper state from diverting the river's waters, the upper riparian is under some obligation to enter into prior consultation. Taubenfeld and Taubenfeld, "Modification of the Human Environment," 133–34. Others who take this view are listed in Springer, *International Law of Pollution*, 149. The support is almost certainly, and exclusively, an environmentally sympathetic reading of *Lac Lanoux*.

17. Springer, *International Law of Pollution*, 149–50, and authorities cited.

18. Lewis, "Atomic Power Safety Steps Approved," sec. 1, p. 36.

19. Handl, "Nuclear Power Siting," 30–39. In fact, the Nordic States Convention, discussed later in the chapter, represents a highly detailed illustration, already in place, of regularized procedures for institutional "conversation" among states regarding mutual environmental problems.

20. Professor Springer points to conventions among the nations fronting Lake Constance (1960) and regarding frontier waters between Austria and Czechoslovakia (1967) as examples of treaties embodying "prior consent" provisions. Springer, *International Law of Pollution*, 150–52. The Austro-Czech agreement provides that "consent may be refused only on serious grounds."

21. Ibid., 69.

22. See "West Germany Announces Projects to Clean Up Environment in East Germany," *International Environment Reporter (BNA)* 12 (1989): 592.

23. 3 R.I.A.A. 1911 (1941).

24. 1949 I.C.J. 4.

25. On the other hand, it appears from the full opinion that Albania's failure to give notice of the mines was its critical misstep (the ICJ so read *Corfu Channel* in *Nicaragua v. United States*), not the fact alone that it set a hazard adrift. See Case Concerning Military and Paramilitary Activities in and Against Nicaragua (Nicar. v. U.S.), 1986 I.C.J. 14, 112. This might suggest, unencouragingly, that a global polluter would escape liability, at least under the *Corfu Channel* rationale, simply by disclosing—by giving notice that it intended to pollute.

26. Springer, *International Law of Pollution*, 157–58; Schneider, *World Public Order of the Environment*, 41–42.

27. See 1984 I.L.C. Rep. 171–92.

28. Christine Gray, *Judicial Remedies in International Law* (Oxford: Clarendon Press, 1987), 69–74.

29. Decision, 16 April 1938, 3 U.N.R.I.A.A. (1949).

30. See J. I. Charney, "Disputes Implicating the Institutional Credibility of the Court: Problems of Non-Appearance, Non-Participation, and Non-Performance," in L. F. Damrosch, *The International Court of Justice at a Crossroads* (Dobbs Ferry, N.Y.: Transnational, 1987), 294, and authorities cited.

31. Nuclear Tests Case (Australia) [1973] I.C.J. 99, 100.

32. Ibid., 106.

33. The Basel Convention does not affect shipment of radioactive wastes, which are under the jurisdiction (such as it is) of the U.N. International Atomic Energy Agency. The Convention was signed by the United States, and ratification by the Senate is expected; however, implementing legislation by Congress will still be required. In the congressional debates, some controversy is shaping up over the health and environmental standards of the recipient country (discussed later in the text).

34. As presently drafted, the convention also suffers because "hazardous" and "waste" are not effectively defined. See David P. Hackett, "An Assessment of the Basel Convention on the Control of Transboundary Movements of Hazardous Wastes and Their Disposal," *American University Journal of International Law and Policy* 5 (1989): 311–12.

35. See Michie v. Great Lakes Steel Division, 495 F.2d 213 (6th Cir. 1974), *cert. denied* 419 U.S. 997 (1974), in which residents of the Windsor area in Canada, in reaction to industrial pollution originating in the Detroit area, chose private remedies in U.S. courts against responsible U.S. corporations, in preference to initiating time-consuming international level remedies.

CHAPTER IV

MANAGING THE GLOBAL COMMONS

1. See U.N. Declaration of Nature, Principle 21, and Restatement (Third) §601, chapter 3, note 2, and text.

2. The net impact (net after allowing for uptake by natural processes) of carbon dioxide-emitting activities is believed to be an annual increase of 3.7 billion metric tons of carbon, presumably in the atmosphere, of which the U.S. share is calculated at 540 million. See World Resources Institute et al., *World Resources 1990–91* (Oxford: Oxford University Press, 1990), 348, 356.

3. Soni, Control of Marine Pollution, 149, citing D. J. Cusine and J. P. Grant, eds., *The Impact of Marine Pollution* (London: Croon Helm, 1980).

4. See World Resources Institute et al., *World Resources 1988–89*, 330.

5. There are narrow situations in which nations have argued for recognition of a quasi-property interest in historically claimable "fishing rights" beyond their territorial waters. See the Fisheries Jurisdiction Case (U.K. v. Iceland), 1974 I.C.J. 3. The U.S. is trying to pressure foreign fleets to recognize U.S control over hake as well as anadromous salmon in the North Pacific beyond the EEZ. See also note, p. 95.

6. See Ian Brownlie, "A Survey of International Customary Rules of Environmental Protection," in Ludwik A. Teclaff and Albert E. Utton, *International Environmental Law*, 5, and Daniel Barstow Magraw, "International Law and Pollution," in *International Law and Pollution*, ed. Daniel B. Magraw (Philadelphia: University of Pennsylvania Press, 1991), 3–29, both pointing to dictum in the *Barcelona Traction* case (Second Phase) (1970). I.C.J. Reports 2, 32.

7. Draft Article 19(d).

8. See Draft Article 14.

9. Indeed, the willingness of states to criminalize environmental misconduct may be impeded by the very vitality of international environmental law: diplomats may want to wait to see what the obligations are that might be criminalized. And conversely, the prospect that breaches of treaty obligations might be criminalized could be dampening the readiness of states to develop clear and thoroughgoing obligations.

10. The LOS would have established an internationally regulated regime for the exploitation of seabed minerals, but the Convention has failed to muster the support of the United States.

11. See the comparison of unilateral versus bilateral and multilateral options in chapter 7.

12. See pp. 52–54.

13. Some 160 nations, including the asserting coastal states, were reported to have acquiesced by 1986. The most common extension of jurisdiction is the EEZ, although there exists a range of other, special extensions, including Canada's pollution control zone and the U.S.'s fisheries management zones.

14. In fact, many commentators point out that it was the prospect that some global agreement was within reach under the LOS, quite likely to the disadvantage of the coastal states, that was a principle catalyst in prompting

the coastal states to seize their EEZ jurisdictions as a way of preempting the LOS conferees, who might have restricted their reach.

15. See the discussion of Canada's pollution control zone, below.

16. See P. Haas, *Saving the Mediterranean* (New York: Columbia University Press, 1990), 128.

17. On the other hand, we shall examine in chapter 6 an important approach to atmospheric and deep oceanic protection, based on a related type of enclosure or "privatization": the establishment of tradable pollution entitlements.

18. Schneider, *World Public Order of the Environment*, 38. See R. T. Scully, "International Regulation of Pollution from Land-Based Sources," *Marine Affairs Journal* 3 (1975): 84–107. The LOS convention proposed a §207 that would operate in this area.

19. Springer, *International Law of Pollution*, 106.

20. Montreal Guidelines for the Protection of the Marine Environment Against Pollution from Land-Based Sources, reprinted in Final Report of the *Ad Hoc* Working Committee of Experts on the Protection of the Marine Environment from Land-Based Sources, U.N. Doc. UNEP/WG.120/3 (1988).

21. The Institute for Energy and Environmental Research asserts that the principal substitute, (HCFCs) are three to five times more dangerous to the ozone layer than government and industry acknowledge. "HCFCs Draw More Fire from Environmentalists," *Chemical Marketing Reporter* 241 (1992): 5. By contrast, UNEP, which supports the substitution, regards HCFCs as more benign ozone depleters than CFCs; moreover, from the perspective of global warming, UNEP estimates that the most deleterious HCFC is only one-twentieth as serious as the worst CFC. See Virginia Kent Dorris, "Drop-in Substitutes for CFCs Are Elusive, Researchers Say," *Engineering News-Record* 223 (1989): 34.

22. For a fuller exposition, see James K. Sebenius, "Crafting a Winning Coalition," in World Resources Institute, *Greenhouse Warming: Negotiating a Global Regime* (Washington, D.C.: World Resources Institute, 1991), 69–98.

23. Two of the scientific contributors to the IPCC report, neither of whom is dismissive of greenhouse risks, have reported that we would pay only a small "penalty" for delaying by ten years a twenty-year transition from the "business-as-usual" to any of the constrained emission scenarios. Schlesinger and Jiang, "Revised Projection of Future Greenhouse Warming," 219. An imminent study suggests that, on certain reasonable assumptions, a policy aimed at only moderate reductions in the near term (1992–2002) is actually

less costly than instituting aggressive reductions immediately. James K. Hammitt et al., "A Sequential-Decision Strategy for Abating Climate Change," *Nature* 357 (1992): 315–18.

24. A recent UNEP study makes even this claim less than 100 percent certain today, indicating that the ozone-depleting agents have at least one virtue—of counteracting greenhouse warming by impairing the heat-retention capacity of the lower stratosphere. See "Stratospheric Ozone Hole Found to Occur in Mid-Hemisphere," *International Environment Reporter (BNA)* 14 (1991): 590.

25. See chapter 5.

26. Under the Montreal Protocol, the exact data are provided to the Secretariat, but are not released publicly in a manner that would reveal details on individual nations and firms. But reportedly, sixty parties were responsible for over 90 percent of the production and consumption of ozone-depleting agents. "Montreal Protocol: Phase-out of CFCs Achieved," *Environmental Policy and Law* 20 (September–October 1990): 134. And it is generally assumed that the dominant national producers were Britain, France, Italy, and the U.S. Richard E. Benedick, *Ozone Diplomacy* (Cambridge, Mass., and London: Harvard University Press, 1991), 68–69. The former three countries accordingly dictated the initial EC position. Ibid., 38.

27. It should be kept in mind, however, that an astonishing 59 percent of anthropogenic carbon dioxide emissions (deforestation aside) is attributable to only five nations: the U.S., the former Soviet Union, China, Japan, and Germany. But if an all-GHG approach is adopted, so that reductions in CO_2 emissions are linked to reductions in other greenhouse agents such as methane, and in deforestation (on the grounds both that the practice produces additional CO_2 and that it impairs the earth's capacity to purge carbon from the atmosphere), then the parties whose cooperation is critical must be augmented to include, at the least, Brazil and India as well. Figures from World Resources Institute et al., *World Resources 1990–91*, 346–47, and table 24.1.

28. Conrad B. MacKerron, "Chemical Firms Search for Ozone-Saving Compounds," *Chemical Engineering News*, January 18, 1988, 22.

29. A project headed by the Union of Concerned Scientists has come out with several energy options that are claimed to be environmentally benign, conservationwise, and economically efficient: it claims that on a $2.7 trillion investment, consumers could save $5 trillion in fuel and electricity costs over a forty-year period, using a 3 percent rate of discount. (At a 7 percent rate, the present value of savings falls to half a trillion dollars.) See Alliance

to Save Energy et al., *America's Energy Choices* (Cambridge, Mass.: Union of Concerned Scientists, 1991), 2 and passim.

30. Alan S. Manne and Richard G. Richels, "CO_2 Emission Limits: An Economic Cost Analysis for the USA," *Energy Journal* 11, no. 2 (1990): 51–85. They project a decrease of roughly 5 percent of total annual macroeconomic consumption, discounted to a present (1990) value at 5 percent. See also idem, "The Costs of Reducing U.S. CO_2 Emissions—Further Sensitivity Analysis," *Energy Journal* 11, no. 4, (1990): 69–78. Nonetheless, the authors maintain that a strong commitment to R&D could conceivably reduce the costs of a carbon constraint, "perhaps by several trillion dollars." *Energy Journal*, 11, no. 2 (1990): 51, 73.

31. See Christopher D. Stone, "Beyond Rio: Insuring Against Global Warming," *American Journal of International Law* 86 (1992): 445–88.

32. The general theory is presented in Christopher D. Stone, *Should Trees Have Standing? Toward Legal Rights for Natural Objects* (Los Altos, Calif.: William Kaufmann, Inc., 1974), and reviewed, with some of the subsequent cases and literature, in idem, *Earth and Other Ethics* (New York: Harper and Row, 1985).

33. See 40 C.F.R. (1990) §§300.600, 300.615(a)(1).

34. United States v. Montrose Chemicals, Dkt. No. CV 90–3122 AAH, D.C.D. Cal. 1990. The case is pending in District Court.

35. See D. Dickson, "Mystery Disease Strikes Europe's Seals," *Science* 241 (1988): 893–94; "Nordsee: Zeichen einer todkranken Natur," *Der Spiegel*, June 6, 1988, 18–28.

36. Stone, *Earth and Other Ethics*.

37. In Germany, the situation varies from state to state; some *Länder* (states), such as Bremen (see §44 Bremisches Naturschutzgesetz) do recognize the right of authorized environmental groups to challenge government action of specified sorts, for example, those affecting forests.

38. The fuller title of the lawsuit is Nochfolgende Ausführungen beziehen sich auf das Antragsverfahren auf Gewährung vorläufigen Rechtsschutzes, das von der Tierspezies der in der Nordsee lebenden Seehunde *gegen* die Bundesrepublik Deutschland (Verwaltungsgericht, Hamburg, August 15, 1988).

39. The court also ruled that the environmental organizations, which had named themselves co-plaintiffs in their own right, had no standing because they were not adversely affected or aggrieved. Decision of the Verwaltungsgericht Hamburg published in *Neue Zeitschrift für Verwaltungsrecht* (NVwZ) 1988 (1057), affirmed by the Oberverwaltungsgericht Hamburg,

November 24, 1988 (OVG Bs VI 49/88), without published opinion November 24, 1988. Independently, shortly after the suit was filed, the scientific evidence shifted focus away from the heavy metals and toward a canine virus as culprit. See "Canine Virus Tied to Seal Deaths," *New York Times*, August 30, 1988, sec. 1, p. 19.

40. See, for example, "Wie Absurd," *Der Spiegel*, September 12, 1988, 71–72. Heinrich von Lersner, president of the German EPA, expressed sympathy for the seals suit in an article in *Der Tagespiel*, October 12, 1989, and in the 1988 *Neue Zeitschrift für Verwaltungsgerecht*, 988–92.

41. Under the Statute of the I.C.J., Art. 65, the World Court will render advisory opinions as provided by the Charter of the United Nations. At present, the U.N. Charter, at Art. 96, authorizes only the General Assembly or Security Council to apply to the Court. The General Assembly has the power to authorize other organs of the U.N. or special agencies to do so, but not states. Presumably, amendments within this framework could provide for guardianships by UNEP, for example, at least to elicit advisory opinions. More far-reaching amendments would be required for fuller guardianship powers by U.N. agencies or NGOs; nations, of course, have the fullest powers, and it has been suggested that a small nation might be prevailed upon to serve as guardian for some aspect of the environment.

42. See Hagen, "Marpol Annex V," 436. See also R. V. Arnaudo, "The Problem of Persistent Plastics and Marine Debris in the Oceans," cited in the Joint Group of Experts on the Scientific Aspects of Marine Pollution, *The State of the Marine Environment* (Oxford: Blackwell Scientific Publications, 1990), 133.

43. See Hagen, "Marpol Annex V," 432–33, citing estimates that perhaps 700,000 birds and 100,000 mammals die from plastic debris annually. The International Union for the Conservation of Nature claimed in 1980 that one millon seabirds were dying in fish nets each year. IUCN *World Conservation Strategy* (Gland, Switz.: IUCN, 1980), sec. 4, par. 5.

44. See chapter 9.

45. See U.N. G.A. Res. 46/215, U.N. GAOR 2d Comm., 46th sess., 79th plen. mtg., U.N. Doc. A/RES/46/215.

46. Convention for the Preservation and Protection of Fur Seals, July 7, 1911, 37 Stat. 1542, 1 *Bevans* 804 (terminated October 23, 1944).

47. Annex V, effective December 31, 1988. Convention for the Prevention of Pollution by Ships, reprinted in *International Environment Reporter Reference File (BNA)* 21: 2301–2400, 2342–43.

48. See J. Cohen, "Was Underwater 'Shot' Harmful to the Whales?" *Science* 252 (1991): 912–14.

49. Most reports anticipated the aftereffects on astronomy as nil, although twenty-five years later Joel Scheraga, "Curbing Pollution in Outer Space," *Technology Review* 89 (1986): 8, claims that West Ford debris was still being accounted for by radio astronomers.

50. See Howard Baker, *Space Debris: Legal and Policy Implications* (Dordrecht and Boston: Nijhoff, 1989), 87–89.

51. Treaty on Principles Governing the Activities of States in the Exploration and Use of Outer Space, Including the Moon and Other Bodies, January 27, 1967, 18 U.S.T. 2410, 610 U.N.T.S. 205. See Carl Christol, *The Modern International Law of Outer Space* (New York: Pergamon Press, 1982), 133–34.

52. See Art. XV.

53. Treaty Banning Nuclear Weapons Tests in the Atmosphere, in Outer Space and Under Water, August 5, 1963, Article I, 14 U.S.T. 1313, 1316, 480 U.N.T.S. 43.

54. Convention on the International Liability for Damage Caused by Space Objects, March 29, 1972, Article II, 24 U.S.T.S. 2389, 2392, 961 U.N.T.S. 187.

55. Canada cleaned up the radioactivity at a cost of $12 million, and made claim against Russia under Article II of the 1972 Convention, which calls for compensation for property damages but does not specifically mention cleanup costs. There was also a claim under customary law. The USSR did not consider cleanup part of the "damage" but paid a portion of the claim with no explanation as to what they were paying for. Scheraga, "Curbing Pollution in Outer Space," 8.

56. Albert Utton, "The Arctic Waters Pollution Prevention Act, and the Right of Self-Protection," in *International Environmental Law*, 41.

CHAPTER V

TREATIES AS ANTIDOTES

1. It bears remembering that treaties, with their preventive stance, are preferable if the problem is real and the benefits of the constraint merit the costs. While traditional ex post (after the fact) damage actions may entail complicated trials, they may at least reduce the risks of overdeterrence: of imposing costs that exceed their benefits.

2. The only example in the pollution field may be the 1954 Convention for the Prevention of Pollution of the Sea by Oil (a Case 4 situation discussed infra); see Springer, *International Law of Pollution*, 157.

3. See Paolo Contini and Peter H. Sand, "Methods to Expedite Environment Protection: International Ecostandards," *American Journal of International Law* 66 (1972): 37–59.

4. On the other hand, majority rule opens up opportunities for the majority to impose upon minorities, so that unanimity, the default collective choice rule of the world order, has its defenders. See Dennis C. Mueller, *Public Choice II* (New York: Cambridge University Press, 1989), 96–111.

5. There is an exception: the World Court (ICJ) has undenied power to issue interim measures of protection while a case is pending, and these often have the same effect as short-term injunctions. See Christine Gray, *Judicial Remedies in International Law* (Oxford and New York: Oxford University Press and Clarendon Press, 1987), 69–74.

6. See John Austin, *The Province of Jurisprudence Determined* (New York: Noonday, 1954), 133, 201.

7. Art. VI (2). The state is not, however, obliged to prosecute. And see Gray, *Judicial Remedies*, 219–22.

8. Wolfgang Friedmann, *The Changing Structure of International Law* (New York: Columbia University Press, 1964), 88–95.

9. S. 984, 102d Cong., 1st sess. (International Pollution Deterrence Act of 1991). See 137 Cong. Rec. S. 5298 (remarks of Sen. Boren, April 25, 1991).

10. Recall that the GATT panel which held that the U.S.'s unilateral boycott of Mexican tuna was a trade violation did not condemn boycotts that effectuated multilateral agreements.

11. See Marc Pallemaerts, "The Politics of Acid Rain Control in Europe," *Environment* 30 (March 1988): 43.

12. See World Resources Institute et al., *World Resources 1988–89*, 167.

13. See "Algae Blooms Signal Trouble," *Ocean* 21 (1988): 69.

14. See Barbara Coleman, Raymond Doetsch, and Roy Sjoblad, "Red Tide: A Recurrent Marine Phenomenon," *Sea Frontiers/Sea Secrets* 32 (1986): 184.

15. "Algae Blooms," 69.

16. See H. E. Landsberg, "Global Climate Trends," in Simon and Kahn, *The Resourceful Earth*, 274.

17. See "Volcanos Can Muddle the Greenhouse," *Science* 245 (1989): 127–28.

18. T. E. Graedel and P. J. Crutzen, "The Changing Atmosphere," *Scientific American* 261 (September 1989): 58–68. As for the recent decline in the rate of methane accumulation, see chapter 1, note 59.

19. See William K. Stevens, "Methane from Guts of Livestock Is New Focus in Global Warming," *New York Times*, November 21, 1989, B7.

20. See Daniel A. Lashof and Dilip R. Ajuhla, "Relative Contributions of Greenhouse Gas Emissions to Global Warming," *Nature* 344 (1990): 529–31. Another, more systems-sensitive approach is to compare GHG-producing *activities* and *technologies* by reference to their entire life-cycle effects. See Rex T. Ellington, Mark Meo, and David E. Baugh, "The Total Greenhouse Warming Forcing of Technical Systems: Analysis for Decision Making," *Journal of the Air & Waste Management Association* 42 (April 1992): 422–28.

21. See "Climate Conflict," *Greenpeace*, July/August 1991, 8.

22. One solution is to negotiate a retroactive baseline date, after which any nation's voluntary reductions will be appropriately credited. See Richard M. Stewart and Jonathana B. Wiener, "A Comprehensive Approach to Climate Change," *American Enterprise*, November/December 1990, 77. See, generally, Peter H. Sand, *Lessons Learned in Global Environmental Governance* (Washington, D.C.: World Resources Institute, 1990).

23. Peter Passell, "Cure for Greenhouse Effect: The Costs Will Be Staggering," *New York Times*, November 19, 1989, 11. See Manne and Richels, *CO$_2$ Emission Limits: An Economic Analysis for the U.S.A.*

24. It is not evident that dark-skinned people would face the same, still largely indeterminate, risk that UV-Bs suppress normal functioning of the immune system. See chapter 1.

25. See UNEP/GEMS, *The Greenhouse Gases* (Nairobi: UNEP, 1987), 22, reporting that, on the basis of an approximately 4° C temperature rise, "three of the most recent model predictions suggest that overall precipitation will increase by between 7 and 11 percent."

26. See UNEP, supra note 25, at 29. A. Yanshin, the Soviet academician and a greenhouse skeptic, pointedly observes from geologic evidence that in past interglacial warming epochs the physico-geographical situation in the USSR (as well as Scandinavia and Central Europe) "was much more favorable than it is today." "Is the Greenhouse Effect Really That Dangerous?" *Scientific World* 34, no. 3 (1990): 14, 15.

27. See World Resources Institute et al., *World Resources 1990–91*, 345, and tables.

28. See Charles Osterberg, "Deep Ocean: The Safest Dump," *New York Times*, June 14, 1989, A19.

29. Andrew Solow, Stephen Polasky, and James Broadus, "On the Measurement of Biological Diversity," *Journal of Environmental Economics and Management* (forthcoming).

30. See "So Much to Save," *The Economist*, June 13, 1992, 94.

31. See Catherine Dold, "To Protect Biodiversity, Expert Says, Save the Dry Land," *New York Times*, April 7, 1992, C4.

32. The issues that arise from efforts to reach a fair division of cooperative surplus are far more complex than indicated in the text; for an excellent discussion, see Brian Barry, *Theories of Justice* 1 (1989): 50–142.

33. See Marlise Simons, "North-South Chasm Is Threatening Search for Environmental Solutions," *New York Times*, March 17, 1992, A5.

34. See, "The Beautiful and the Dammed," *Economist* 322 (1992): 93–95.

35. See Charles S. Pearson, *International Marine Environment Policy* (Baltimore: Johns Hopkins University Press, 1975), 58–59 (indeterminate effect of injecting environmental values into LOS negotiations).

36. See Dennis C. Mueller, *Public Choice II*, 58–59; David A. Wirth, "The World Bank and the Environment," *Environment* 28, no. 10 (1986): 33–34.

37. Although that linkage is ordinarily urged in the course of bilateral, as opposed to large-scale multilateral negotiations.

38. See, generally, Robert W. Hahn and Kenneth R. Richards, "The Internationalization of Environmental Regulation," *Harvard International Law Journal* 30 (1989): 421–40.

39. There were other technical, scientific, and economic reasons that hampered the nitrogen oxide agreements, but the United States and other nations eventually signed the protocol to the 1979 Convention on Long-Range Transboundary Air Pollution Concerning the Control of Emissions of Nitrogen Oxides or their Transboundary Fluxes, October 31, 1988. Reprinted in *International Legal Materials* 28 (1989): 212–30.

40. Doubts have since crept in, and HCFCs themselves now face a rapid phasing out in favor of a new generation of imperfectly understood substitutes. See chapter 1.

41. Jacobson and Kay underscore the value of normative pronouncements in advancing policy-coordinating. Citing the Stockholm Declaration, they observe: "Although declarations of principle do not have the same legally binding consequences as conventions or treaties, they nonetheless often have a substantial effect on national policies, and they enjoy the substantial advantage of being easier to adopt. Officials in government can point to such declarations and argue that their state should adhere to the norms that have been enunciated." Harold K. Jacobson and David A. Kay, "A Framework for Analysis," in *Environmental Protection: The International Dimension*, ed. Kay and Jacobson (Totowa, N.J.: Allanheld, Osmun & Co., 1983), 16.

42. Noel Grove, "An Atmosphere of Uncertainty," *National Geographic* 171 (April 1987), 522.

43. Pearson, *International Marine Environmental Policy*, 45. Pearson proceeds to give examples of the institutional by-products of the conference, including establishment of the International Working Group on Marine Pollution (IWG).

CHAPTER VI
THE ECONOMIST'S PRESCRIPTIONS

1. India's demand was backed up by a UNEP survey that concluded that India would lose $2 billion by replacing existing CFC plants with substitute facilities. This figure is conservative because CFC substitutes require more energy to utilize. "Technology: Change to CFC-Substitute Production Would Cost India $2 Billion, UNEP Says," *International Environment Reporter (BNA)* 13 (1990): 171. Nonetheless, the Indian cabinet agreed to ratify after negotiators made two concessions to developing countries. First, they offered developing countries a grace period to comply with the requirement. Second, they offered financial and technical assistance to expedite the phaseout. "India: Cabinet Agrees to Ratify Montreal Protocol on Substances Depleting Earth's Ozone Layer," *International Environment Reporter (BNA)* 14 (1991): 188–89. The protocol established a $160 million fund to accelerate developing countries' substitution, with an extra $80 million for new signatories (principally, China and India). "Air Pollution: Parties to Montreal Protocol Agree to Phase Out CFCs, Help Developing Nations," *International Environment Reporter (BNA)* 13 (1990): 275.

2. Among the cargo that passes between Basel and Rotterdam every year was at one point 3,150 tons of chromium, 1,520 tons of copper, 12,300 tons of zinc, 70 tons of mercury, and 350 tons of arsenic. See A. Rest, "A Decision against France?—The Rhine Pollution," *Environmental Policy & Law* 5 (1979): 85–89.

3. See William J. Baumol and Wallace E. Oates, *The Theory of Environmental Policy*, 2d ed. (Cambridge, U.K.: Cambridge University Press, 1988), 281. This volume is highly recommended for those who wish a deeper grounding in environmental policy and economics.

4. World Resources Institute et al., *World Resources 1987* (New York: Basic Books, 1987), 188. All told, the proposed actions will cost $90 million. France and Germany will pay 30 percent each, the Netherlands 34 percent, and Switzerland 6 percent. "The Rhine: Protocol to Curb Saline

Discharges into Rhine Approved by Five Nations," *International Environment Reporter (BNA)* 14 (1991): 545.

5. Article 2 of the sulfur dioxide protocol to the Convention on Long Range Transboundary Air Pollution (LRTAP), signed by twenty-one states from East and West in Helsinki on July 9, 1985, provides for the parties to reduce their national annual sulfur emissions or their transboundary fluxes by at least 30 percent as soon as possible and at the latest by 1993, using 1980 levels as the basis for calculation of reductions. However, most industrialized nations had already begun SO_2 reductions in the 1970s, with emissions down 20 to 60 percent in the 1975–84 period. See World Resources Institute et al., *World Resources 1988–89*, 165. The signatories simply ratified the direction in which the industrialized nations were heading. The climate change negotiators face the considerably more difficult challenge of changing direction.

6. There are of course various techniques with which to address these "equity" problems and avoid a diplomatic standoff, for example, differential schedules for compliance, and even side payments as an inducement to sign. See, generally, Daniel B. Magraw, "Legal Treatment of Developing Countries: Differential, Contextual, and Absolute Norms," *Colorado Journal of International Environmental Law and Policy* 1 (1990): 69–99.

7. Marginal costs of reduction efforts are presumed to increase with units eliminated—that is, typically, the first 10 percent of a pollutant is the cheapest to remove; the last 10 percent may be prohibitively expensive.

8. See Bruce A. Ackerman et al., *The Uncertain Search for Environmental Quality* (New York: The Free Press, 1974), 224–27.

9. The tax receipts are generally directed to the general treasury, but this need not be so. Since 1970 the Netherlands have invoked a pollution tax with revenues earmarked for the construction of public waste treatment facilities. In their report of Netherlands cleanup efforts, Gjalt Huppes and Robert A. Kagan suggest that this earmarking has reduced resistance by industry. See Gjalt Huppes and Robert A. Kagan, "Market Oriented Regulation of Environmental Problems in the Netherlands," *Law & Policy* 11 (1989): 215–39.

10. See 15 U.S.C. §2002, 2003, 2008 (1982); 49 C.F.R. §531.5 (Rev. Oct. 1, 1991).

11. See Huppes and Kagan, "Regulation of Environmental Problems." The authors present several conjectures as to why the one regime has worked better than the other.

12. See Donald T. Rocen and David C. Green, "New Tax on Ozone Depleting Chemicals Has Far-Reaching Consequences," *Journal of Taxation* 72

(1990): 282–85. To convey a sense of scale, CFC producer prices have been averaging about $1 a pound; taxes of most common compounds started at $1.37 in 1990, to escalate to $5 a pound by the end of the century. John Holusha, "Ozone Issue: Economics of a Ban," *New York Times*, January 11, 1990, D1.

13. William A. Nitze appears to be an advocate of this mode, suggesting that an international carbon tax would be made more palatable by allowing each country to keep the revenues raised for its own domestic purposes. See William A. Nitze, *The Greenhouse Effect: Formulating a Convention* (London: Royal Institute of International Affairs, 1990), 52. Readers wishing to pursue the tax options are referred to James M. Porteba, "Tax Policy to Combat Global Warming: On Designing a Carbon Tax," and John Whalley and Randle Wigle, "The International Incidence of Carbon Taxes," both in *Global Warming: Economic Policy Responses*, ed. Roger Dornbusch and James M. Poterba, (Cambridge, Mass.: MIT Press, 1991).

14. Michael Grubb, "The Greenhouse Effect: Negotiating Changes," *International Affairs* 66 (1990): 67, 70–71, 79–81.

15. Daniel A. Lashof and Dennis A. Tirpak, eds., *Policy Options of Stabilizing Global Climate* (Washington, D.C.: EPA Office of Policy, Planning, and Evaluation, 1989), 771.

16. See Joshua M. Epstein and Raj Gupta, *Controlling the Greenhouse Effect: Five Global Regimes Compared* (Washington, D.C.: Brookings Institution, 1990), 18–20.

17. See "Battle Looming over Energy Tax Commission Expected to Propose This Month," *International Environment Reporter (BNA)* 14 (1991): 481. France, as the heaviest user of nuclear energy, is reportedly willing to put a heavy penalty on carbon use.

18. Stephen Seidel and Dale L. Keyes, *Can We Delay a Greenhouse Warming?* (Washington, D.C.: EPA, 1983), 4–31.

19. See C. Holden, "Greenhouse Gas Tax," *Science* 255 (January 10, 1992): 154, reporting on Office of Environmental Analysis, Department of Energy, *Limiting Net Greenhouse Gas Emissions in the United States* (December 1991). Remember that the loss to people qua consumers might be offset by a benefit to them qua taxpayers, to the extent the government were to offset other forms of taxation by the carbon revenues.

20. Grubb, *Greenhouse Effect*, 80–81. This is a criticism of taxes in the second mode, in which revenues are collected and applied by an international authority.

21. Bruce A. Ackerman and Richard B. Stewart, "Reforming Environmental Law," *Stanford Law Review* 37 (1985): 1333–65. The EPA-proposed

regulations on CFCs discusses the auction option and appears to reject it, although not on persuasive grounds.

22. For example, Baumol and Oates demonstrate that all other things being equal, the steeper the slope of the marginal benefits function, the less distortion from regulatory misjudgment about the cost function will result under a permit (quantity fixing) system than under a fee (price fixing) equivalent. Conversely, the steeper the slope of marginal control costs, the larger the opportunities for distortion under the permit system. Baumol and Oates, *Theory of Environmental Policy*, 64–66. Is anyone really bold enough to declare which is the most likely direction of error in the GHG context?

23. The classic article is Martin L. Weitzman, "Prices vs. Quantities," *Review of Economic Studies* 41 (1974): 477–91.

24. Conversely, if an agreed-upon cost constraint emerges at the political level, fees dominate.

25. See, generally, Baumol and Oates, *Theory of Environmental Policy*, chapter 12.

26. See Robert W. Hahn and Gordon L. Hester, "Where Did All the Markets Go? An Analysis of EPA's Emissions Trading Program," *Yale Journal of Regulation* 6 (1989): 109.

27. See "Pollutants Reined by Market Rules," *Insight*, July 3, 1989, 9–17.

28. Rex Rhein, "Environmental Incentives Get a Market Outlook," *Chemical Engineering* 99 (1992): 62.

29. "Futures Trading: CFTC Approves CBT Proposal for Sulfur Dioxide Emission Futures," *Daily Report for Executives (BNA)*, April 22, 1992, Regulation, Economics, and Law section, A10.

30. Barbara Durr, "Deal Sets Up Trade in Pollution Permits," *Financial Times*, May 13, 1992, 7.

31. For analysis of a whole range of legal and institutional issues, see Jonathan Green and Philippe Sands, "Establishing an International System for Trading Pollution Rights," *International Environment Reporter (BNA)* 15 (1992): 80.

32. Arthur H. Westing, "Law of the Air," *Environment* 31 (1989): 187, would allot nations one "CO_2 release chit" per square kilometer of territory, each chit allowing one 133 millionth of the total permissible emissions.

33. Epstein and Gupta, *Controlling the Greenhouse Effect*, 18–20. I cannot argue the point adequately in this space, but population does not seem normatively superior to other bases—to grandfathering, for example, on some of the same bases that support recognizing the title of an adverse possessor, or to a need- or capabilities-regarding index, along lines discussed

by Amartya Sen. See Amartya K. Sen, *Resources, Values and Development* (Cambridge, Mass.: Harvard University Press, 1984), 307–24 ("Rights and Capabilities"), and 325–45 ("Poor, Relatively Speaking").

34. Article II of the London Amendments to the Montreal Protocol provides that a party may transfer to another portions of a party's allowed share of controlled substance pollution, provided that the combined pollution of the two trading nations does not exceed prescribed limits. Montreal Protocol on Substances That Deplete the Ozone Layer: Composite Text of Protocol as Adjusted and Amended in June 1990 in London, reprinted in *International Environment Reporter Reference File (BNA)* 21: 3151–94, 3151–53.

35. See "Paying Off Fishermen May Restore Salmon," *Wall Street Journal*, August 25, 1992, B1. The background is to be found in NASCO, *Report of the Activities of the North Atlantic Salmon Conservation Organization in 1989–90* (Edinburgh, Scotland: NASCO, 1991), 7.

36. See Mark Sagoff, *The Economy of the Earth: Philosophy, Law, and the Environment* (Cambridge, U.K.: Cambridge University Press, 1988), 97; see review, Christopher D. Stone, *Environmental Ethics* 10 (1988): 363.

37. Prior to 1992, the fund paid out $12,000 over a five-year period in indemnity for lost cattle. See William Kronholm, "Conservation Group Offers Bonus to Ranchers Hosting Wolf Packs," *Laramie Daily Boomerang*, June 14, 1992, 12.

38. Agnus M. Thurmer, Jr., "Yellowstone Wolves May be Worth $30 M," *Jackson Hole News*, June 24, 1992, 19A. An astonishing 26 percent of people polled in a nationwide survey said they had been to Yellowstone; 88 percent wanted to go. Sixty-three percent of the visitors reportedly would pay $5 to see the wolves; 13 percent say they would pay $200. The story does not indicate whether the pollster caught the reduced worth of the park with fewer elk, etc.

39. See Amartya K. Sen, *On Ethics and Economics* (Oxford: Basil Blackwell, 1987).

CHAPTER VII

MEDICATING THE EARTH

1. See L. R. Ember et al., "Tending the Global Commons," *Chemical and Engineering News* 64 (November 24, 1986): 14–15, 49.

2. See Bernard H. Oxman, "The Two Conferences," in *The Law of the Sea*, ed. Bernard H. Oxman et al. (San Francisco: Institute for Contemporary Studies, 1983), 127, 135–36.

3. Convention for the Administration of Navigation on the Rhine, August 15, 1804, 1 de Martens 261.

4. Administering the proposed labyrinthine structure would have assured continued employment to the cadre of LOS draftsmen, many of whom had been doing little else for decades and otherwise faced uncertain careers, but would also thereby have dissipated most of the hoped-for royalty revenues from the contemplated mineral production. See Arvid Pardo, "An Opportunity Lost," in *Law of the Sea*, 13, 23. Re the ideological motivation, see Joseph S. Nye, Jr., "Political Lessons of the New Law of the Sea Regime," in ibid., 113, 118.

5. The ostensible connection was that Israel had interfered with telecommunications in southern Lebanon; the United States succeeded in quelling the movement only by threatening to pull out of the ITU if the Arab proposal succeeded. See "Arab Nations Abandoned Efforts . . .," UPI Report, October 21, 1982, AM cycle.

6. See Abram Chayes, "International Institutions for the Environment," in *Law, Institutions and the Global Environment*, ed. John L. Hargrove (Dobbs Ferry, N.Y.: Oceana Publications, 1972). Another pessimistic account is Zdenek J. Slouka, "International Environmental Controls in the Scientific Age," in ibid.

7. On regime theory, generally, see *International Regimes* ed. Stephen D. Krasner (Ithaca, N.Y.: Cornell University Press, 1983); Oran R. Young, *International Cooperation: Building Regimes for Natural Resources and the Environment* (Ithaca, N.Y.: Cornell University Press, 1989). A recent survey of the theoretical work is S. Haggard and B. A. Simmons, "Theories of International Regimes," *International Organization* 41 (1987): 491–517; Oran R. Young, "International Regimes: Toward a New Theory of Institutions," *World Politics* 39 (1986): 104–22.

8. Thomas C. Schelling, "Climatic Change: Implications for Welfare and Policy," in *Changing Climate*, Board on Atmospheric Sciences and Climate, Commission on Physical Sciences, Mathematics, and Resources, National Research Council (Washington, D.C.: National Academy Press, 1982), 449–82.

9. Ibid., 450.

10. The notion of the "second best" is that the next best alternative to an ideal but ultimately unachievable solution is not necessarily a watered-down version of the ideal, for example, a multinational permit trading or tax regime that is practically toothless or leaves out major emitters. The circumstances that make the ideal unrealizable may make it imperative to rerank all

options from the start; the most viable of the new set of alternatives may be an option quite different from the best choice in the original ranking. See, generally, Richard G. Lipsey and Kelvin Lancaster, "The General Theory of Second-Best," *Review of Economic Studies* 24 (1956): 11.

11. On the other hand, Daniel Barstow Magraw presents some appealing reasons why relatively open-textured terms can play a particularly valuable role in the process of international lawmaking. See Daniel Barstow Magraw, "Legal Treatment of Developing Countries: Differential, Contextual, and Absolute Norms," *Colorado Journal of International Law and Policy* 1 (1990): 69–99.

12. "Helsinki Rules on the Uses of the Waters of International Rivers," chapter 2, article 4, in The International Law Association, *Report of the Fifty-Second Conference* (London: International Law Association, 1966), 486–87.

13. International Law Commission, *Draft Articles of the Law of the Non-Navigational Uses of International Watercourses*, art. 7, U.N. Doc. A/CN.4/L.463/Add.4 (1991).

14. Although long favored by academics, marketable permit water use systems have been instituted in the U.S. but generally with success only when done on a relatively small scale. An impediment to instituting the system is said to be the third-party effects of a trade. Picture three parties (farmers or nations) all lined up along a river, Upper (*U*), Middle (*M*), and Lower (*L*); *M* has been using its water for a nonconsumptive use, say hydroelectric (so that it eventually restores to the river's flow virtually all of the water it "uses"); then a new nonhydroelectric source of power becomes viable for *M*—a coal fired plant with excess capacity. Not needing as much river flow, *M* sells its water use entitlement to *U* for *U*'s consumptive use. This *U-M* sale affects *L*, since *L*'s flow will be less than it was. U.S. courts are inclined to say that *L* has "a property right in the water arrangement," and has therefore been empowered to draw the courts (or agencies) back into the fray as an interested third party. My own impression is the friction could be overcome with more imaginative entitlements, if there were only the will and resourcefulness.

15. World Resources Institute et al., *World Resources 1988–89*, 102.

16. Christopher D. Stone, "The Place of Enterprise Liability in the Control of Corporate Conduct," *Yale Law Journal* 90 (1980): 1–77.

17. In ibid.

18. The leading proponent of deploying standards in the international arena has been Peter H. Sand, "Innovations in International Environmental Guidance," *Environment* 32 (November 1990): 16–20, 40–43, reporting

how the International Labor Organization (ILO), going back to the 1920s, has monitored many multilateral conventions that invoke health standards; its "auditing system" enlists trade unions and employers' associations.

19. See Andronico O. Adede, "Overview of Legal and Technical Aspects of Nuclear Accident Pollution," in *International Law and Pollution*.

20. See Paoli Contini and Peter H. Sand, "Methods to Expedite Environmental Protection: International Ecostandards," *American Journal of International Law* 66 (1972): 48–49.

21. See "EC Agriculture Ministers Fail to Agree on Community-Wide Register for Pesticides," *International Environment Reporter (BNA)* 14 (1991): 296.

22. René Dubos, *Reason Awake: Science for Man* (New York: Columbia University Press, 1970), 195.

23. Ibid. Unanticipated benefits of environmental standards are common. Allied Signal Company reports that a new polymer-membrane process designed to scrub coal can capture sulfur dioxide in the form of marketable sulfuric acid (and other by-products). See Amal Kumar Naj, "Turning Smokestack Gas into Useful Products," *Wall Street Journal*, April 17, 1992, B1.

24. See Goldie, "International Maritime Environmental Law Today—An Appraisal," in *Who Protects the Oceans*, ed. Hargrove.

25. Nordic Convention on the Protection of the Environment, February 19, 1974, described in "Current Legal Developments," *International and Comparative Law Quarterly* 23 (1974): 886–87.

26. Tennessee Valley Authority v. Hill, 437 U.S. 153 (1978). Following the decision in *Hill*, the ESA was amended to allow a special exempting procedure, which it was contemplated would be invoked to exempt the Tellico project. But in the face of growing misgivings about the project for a number of reasons, the exemption procedure was not invoked. Congress subsequently intervened to specifically direct that the dam be completed. The reservoir was filled in 1980; some snail darters were transplanted. Later, another indigenous population was discovered in a different stream. See note in John E. Bonine and Thomas O. McGarity, *The Law of Environmental Protection* (St. Paul, Minn.: West, 1992), 229.

27. Pierre B. Crosson and Norman J. Rosenberg, "Strategies for Agriculture," *Scientific American* 261 (September 1989): 128–35.

28. This is the view of former governor Walter Hickel, and I believe he is exactly right. See John Balzar, "Industry's Feeding Frenzy Perils Richest

U.S. Fishery," *Los Angeles Times*, June 20, 1992, A1 (part of a three-part series on the Alaskan fishery that cries out for a broad and continuing audience).

29. See "A Sustainable Stock of Fishermen," *The Economist*, January 19, 1991, 17 (discussing failure of EC's Common Fisheries Policy to halt overfishing).

30. See chapter 2.

31. For an indication of how much the local policies of timber exporting countries continue to frustrate their own presumed interests in sustained-yield forestry, see Jeffrey R. Vincent, "The Tropical Timber Trade and Sustainable Development," *Science* 256 (1992): 1651–55.

32. See Donald G. McNeil, Jr., "How Most of the Public Forests Are Sold to Loggers at a Loss," *New York Times*, November 3, 1991, sec. 4, p. 2.

33. See editorial, "Alaska Travesty," *Los Angeles Times*, May 7, 1986, sec. 2, p. 4.

34. Richard Nelson, review of *The Good Rain* by Timothy Egan, *Los Angeles Times*, July 15, 1990, Book Review section, 3.

35. Editorial, "Fire Sale in a Rain Forest," *Los Angeles Times*, August 9, 1988, B6.

36. In 1990 a House-Senate Conference Committee approved a bill reducing present access to logging from 1.7 million to 1.4 million (of the park's 17 million) acres and requires the Service to prepare for sale only enough timber as will meet "market demand" (whatever that means). Associated Press, "Senate Votes to Preserve Part of Alaska Rain Forest," *Los Angeles Times*, June 14, 1990, A22.

37. In the U.S., Long John Silver's seafood chain canceled a $9 million contract for frozen Icelandic cod. Tengelmann supermarket chain of Germany canceled a $3 million contract for Icelandic shrimp; the Boston School Committee was pressured out of $250,000 of Icelandic fish. *Greenpeace* 14 (January–February 1989).

38. A particularly strong hand has been played by Bruce Rich and the Environmental Defense Fund; see *EDF Newsletter*, March 1987. Bruce Rich, "The Multilateral Development Banks, Environmental Policy, and the United States," *Ecology Law Quarterly* 12 (1985): 681–745.

39. See, for example, for Bremen, §44 Bremisches Naturschutzgesetz; for Hessen, §36 Hessisches Naturschutzgesetz.

40. Resolution 1296. "Arrangements for Consultation with Non-govern-

mental Organizations" (1520th plen. mtg., 23 May 1968), Official Records of the Economic and Social Council, 44th sess., supp. no. 1 (E/4548).

41. See *Drafts: Agenda 21, Rio Declaration, Forest Principles* (New York: United Nations, 1992), Agenda 21, chapter 38.

CHAPTER VIII
TAKING OUT CALAMITY INSURANCE

1. See Vanuatu, draft annex relating to Article 23 (Insurance) for inclusion in the revised single text relating to mechanisms (U.N. Doc. A/AC.237/ WG.II/Misc.13) submitted by the cochairmen of Working Group II, International Negotiating Committee for a Framework Convention on Climate Change, 4th sess., Geneva, December 9–20, 1991, discussed above. U.N. Doc. A/AC.237/WG.II/CRP.8 (1991). See "Small Island Nations to Seek Accord on Creating Insurance Pool to Cover Risk," *International Environment Reporter (BNA)* 14 (1991): 561–62. As I will point out below, the kind of insurance nations in the island states' position might reasonably prefer is more akin to health than to property insurance. "Insurance Against Climate Disasters Taken Up by Climate Change Negotiators," *International Environment Reporter (BNA)* 14 (1991): 675. The final appearance is in Article 4, sec. 8, of the Climate Change Convention, *International Environment Reporter (BNA)* 21: 3904.

2. Stephen H. Schneider, *Global Warming* (San Francisco: Sierra Club Books, 1989), 283–84. Compare David Pearce, *Blueprint 2: Greening the World Economy* (London: Earthscan, 1991), 22: "The way to behave in the face of scientific and economic uncertainty about the greenhouse effect is to adopt fairly restrictive measures."

3. William D. Ruckelshaus, "Toward a Sustainable World," *Scientific American* 261 (September 1989): 166–75. Ruckelshaus continues: "As long as we are going to pay premiums, we might as well pay them in ways that will yield dividends in the form of greater efficiency, improved human health or more widely distributed prosperity. If we turn out to be wrong on greenhouse warming . . . we may still retain the dividend benefits. In any case, no one complains to the insurance company when the disaster does not strike." True; but one may well complain to oneself about wasted premiums, particularly if one discovers they were higher than the risk warranted.

4. One might rejoin that this is simply a case of the hypothetical homeowner valuing the house at greater than the market-measured replacement

cost; but (while I cannot argue the point fully here) I think there is more to the analysis than that.

5. See chapter 1.

6. In this sense, an authentic optimist might contend that the continued use of GHGs is not a pure risk but a speculative one, like an investment.

7. Intergovernmental Panel on Climate Change, Policymakers Summary of the Potential Impacts of Climate Change: Report from Working Group II to IPCC, 28. The combined present populations of these micronations (total area = 402 sqare miles) is only 276,113, with a combined gross product of U.S. $92 million. See, generally, *World Almanac 1990* (New York: Pharos Books, 1989).

8. North American Reinsurance Corp., Sigma, *Natural Catastrophe and Major Losses 1970–1989: Increasing Catastrophe Losses from Forces of Nature in the 1980s* (Zurich: Swiss Reinsurance Co., 1990), 4.

9. See, generally, Charles Weiss, Jr., "Can Market Mechanisms Ameliorate the Effects of Long-Term Climate Change?" *Climatic Change* 15 (1989): 299–307.

10. In other words, suppose that, notwithstanding the impediments to civil liability for climate-change-driven damages discussed in chapter 3, some liability claims are successfully made. Public policy permitting, parties subject to those liability claims might be allowed to insure against them.

11. See Michael Wilford, *Insuring Against the Consequences of Sea-Level Rise* (London: Centre for International Environmental Law, King's College, 1991) (arguing for an insurance pool funded by the developed nations to cover losses from sea-level rise).

12. See Carolyn Aldred, "Insurers Fear Global Warming to Hike Losses," *Business Insurance*, February 12, 1990, 1.

13. See Wilford, *Insuring Against Sea-Level Rise*, 5. Each of these responses, however, will only intensify the pressures for adverse selection.

14. Indeed, even publicly supplemented insurance, discussed later in the text, may well be inadequate for many of these risks; to the extent that insurance is defective, the argument for more conventionally discussed risk-avoidance measures, such as mandatory emission reductions, is strengthened.

15. Weiss, *Market Mechanisms*, 305.

16. Kenneth S. Abraham, *Distributing Risk: Insurance, Legal Theory and Public Policy* (New Haven: Yale University Press, 1986), 90.

17. Property insurance, like life insurance, provides cash payments contingent on some occurrence, for example the burning of a house, and leaves

the insured free to use the proceeds for whatever he or she chooses: you don't have to rebuild your house, you can take a holiday cruise. By contrast, health insurance payments are restricted to reimbursement for approved services. See "Optimal Insurance and Generalized Deductions," in Kenneth Arrow, *Individual Choice under Certainty and Uncertainty* (Cambridge, Mass.: Belknap Press, 1984), 212–16.

18. International Convention on the Establishment of an International Fund for Compensation for Oil Pollution Damage, December 18, 1971, *International Legal Materials* 11 (1972): 284–302, with Protocol, November 19, 1976, *International Legal Materials* 16 (1977): 617–21.

19. The Convention Supplementary to the Paris Convention of July 29, 1960, on Third Party Liability in the Field of Nuclear Energy, January 31, 1963, modified by Additional Protocol, January 28, 1964, International Atomic Energy Agency, *International Conventions on Civil Liability for Nuclear Damage* 43 (1976). See also Wilford, *Insuring Against Sea-Level Rise*, 9–10.

20. See Bob Drogin, "Amid Devastation, Ormoc Buries Its Dead," *Los Angeles Times*, November 8, 1991, A6.

21. Edited by Paul Halstead and John O'Shea (Cambridge, U.K.: Cambridge University Press, 1989), 3–4.

22. International Union for Conservation of Nature and Natural Resources, *World Conservation Strategy* (Gland, Switzerland: IUCN, 1980), part 17; Donna K. H. Walters and Tamara Jones, "Eastern German Seed Bank a Living Legacy for the World," *Los Angeles Times*, September 24, 1991, H3.

23. Lester Lave, "Mitigating Strategies for Carbon Dioxide Problems," *American Economic Review* 72 (Papers and Proceedings issue, 1982): 260.

24. See The Robert T. Stafford Disaster Relief and Emergency Assistance Act, 42 U.S.C. §§5121–5201 (1988).

CHAPTER IX

PAYING THE BILLS

1. How the world would differ if we revoked the deductibility of charitable tax contributions is actually quite uncertain. See Jeff Strnad, "The Charitable Contribution Deduction: A Politico-Economic Analysis," in *The Economics of Nonprofit Institutions*, ed. Susan Rose-Ackerman (Oxford: Oxford University Press, 1986), 265–96.

2. See J. Stoessinger et al., *Financing the United Nations System* (Washington, D.C.: Brookings Institution, 1964), 80. As for the legality of withholding, see R. W. Nelson, "International Law and U.S. Withholding of Pay-

ments to International Organizations," *American Journal of International Law* 80 (1986): 973–83.

3. See "U.S. Parts Ways with UNESCO," *New York Times*, December 23, 1984, sec. 4, p. 1. B. Drummond Ayres, Jr., "U.S. Affirms Plan to Leave UNESCO at End of Month," *New York Times*, December 20, 1984, A1.

4. On the merits, the U.S. was probably right to have coerced reform by withdrawing support from UNESCO. But viewed procedurally, it is a tactic unavailable to most of the world and fuels charges that most Americans find genuinely puzzling: that the U.N. is a tool of U.S. foreign policy.

5. See Articles 20, 21, 1992, United Nations Framework Convention on Biological Diversity, reprinted in *International Environment Reporter Reference File (BNA)*, 21: 4001.

6. Most notably in Grenville Clark and Louis Sohn, *World Peace through World Law* (Cambridge, Mass.: Harvard University Press, 1966); see Stoessinger et al., *Financing the United Nations System*, 280–81.

7. See "Gandhi Calls for $18 Billion Fund to Fight Pollution of Atmosphere," *Los Angeles Times*, September 6, 1989, part 1, p. 8. This is a modest tax in comparison with that envisioned by Clark and Sohn in *World Peace through World Law*: under their plan the U.N. would assign annual revenue quotas, not to exceed 2 percent of estimated gross world product, to the people of each member state. Stoessinger et al., *Financing the United Nations System*, 280.

8. In April 1989 Gro Harlem Bruntdland stated that Norway was prepared to contribute U.S. $90 million into an international climate fund under the auspices of the United Nations, conditioned on matching support from other industrial nations. There was little detail—"environmental organizations dismissed Bruntdland's program as vague and self-promoting"—and nothing ever came of it. See "Norway's Bruntdland Proposes International Climate Fund," *Reuter Library Report*, April 28, 1989.

9. See "International Pollution Tax Proposed to Support Conservation, Climate Research," *International Environment Reporter (BNA)* 13 (1990): 61. Considering that annual carbon emissions run to 700 million tons, it appears that de Collor did not include emissions from, among other activities, deforestation.

10. *Policy Options for Stabilizing Global Climate*, ed. Daniel A. Lashof and Dennis A. Tirpak (Washington, D.C.: EPA, 1991), chapter 7, p. 53, table 7–11. A tax on all domestic carbon of $15 per ton is variously predicted to result in a reduction of 27 to 182 million tons per year. (Five dollars a ton is equivalent to one cent per gallon of gasoline.)

11. Stephen Seidel and Dale L. Keyes, *Can We Delay a Greenhouse Warming?* (Washington, D.C.: EPA, 1983). By the year 2100 (if one plays those games), the tripling of carbon taxes, with its enormous risk to world economies, is projected to make the world only 1.3° C cooler. Ibid., chapter 4, pp. 27–31.

12. In the Oslo Convention area alone, where some tabulations are attempted, 6 million tons were deposited in one form or another. See World Resources Institute et al., *World Resources 1990–91*, table 23.5. (The Oslo Convention applies to ship and aircraft dumping in the North Atlantic and Arctic bounded by 26°N, 42°W, and 51°E. Internal waters, harbors, and estuaries are not covered by the convention.)

13. An excellent brief treatment remains Charles S. Pearson, "Extracting Rent from Ocean Resources," *Ocean Development and International Law Journal* (1973) 1: 221–37.

14. Our calculations are based on the conclusions of the Committee on Science, Engineering, and Public Policy, *Policy Implications of Greenhouse Warming* (Washington, D.C.: National Academy Press, 1992), 8, table 2.1 (drawing on a study of U.S. Department of Energy), and the World Resources Institute et al., *World Resources 1990–91*, 346–47, table 24.1.

15. As of 1980, roughly 160 million metric tons of SO_2 were emitted annually. Geoffrey Lean et al., *Atlas of the Environment* (New York: Prentice-Hall, 1990), 91.

16. One leading observer of the process, the late Harvey Levin, suggested that the present value of positions was aggravated by a shortage artificially created "by international administrative sluggishness and incompetence." See "Annenberg Speakers Push 'Market Solution' for Spectrum Allocation," *Communications Daily* 11 (May 1, 1991): 2.

17. J.E.S. Fawcett, *Outer Space* (Oxford: Oxford University Press, 1984), 88.

18. The text is set forth at *Journal of Space Law* 6 (1979): 193–96.

19. Peter N. Spotts, "World Radio Conference to Allocate Frequencies," *Christian Science Monitor*, February 3, 1992, 7.

20. See "Annenberg Speakers Push 'Market Solution' for Spectrum Allocation." The move by Tonga has sparked controversy, if only for its brazenness; see "Intelsat Assails Satellite Orbit Grab by Island of Tonga," *Reuter Business Report*, August 28, 1990, BC Cycle.

21. "Sandoz to Contribute to WWF Project to Restore Rhine River's Flora, Fauna," *International Environment Reporter (BNA)* 11 (1988): 108. Another prospective source of revenues for the fund would be user charges for

various services of commons-connected world agencies, such as fees for uses of world weather services.

22. This is only roughly how an ideal (Pigovian) tax would work. More exactly, the tax would be set at the level that raised the long-run marginal costs of abating the pollution to the point where it equated the social benefit of further abatement. See the discussion of the effluent tax in chapter 6.

23. See World Resources Institute et al., *World Resources 1990–91*, 186–88.

24. In general, it can be said that the various fund proposals are vague either on source or application or both. See Special Report, "Scientists from 48 Countries View Tax on Fossil Fuel Consumption as a Way of Helping Pay for Action Plan to Safeguard Global Atmosphere," *International Environment Reporter (BNA)* 11 (1988): 414–18. William A. Nitze, in advocating a climate convention, asks that it include a fund "to meet all or part of the hard currency costs of preparing and updating the developing countries' national plans" to create energy-efficient systems, etc. Nitze, "A Proposed Structure for an International Convention on Climate Change," *Science* 249 (1990): 607–8. The source of the funds is not identified. William Ruckelshaus, in the special edition of *Scientific American*, mentions a "climate protection tax" but apparently only as a means of constraining carbon use; application of tax revenues is not mentioned. William Ruckelshaus, "Toward a Sustainable World," *Scientific American* 261 (September 1989): 172. In 1989, at Prime Minister Margaret Thatcher's Greenhouse Seminar, Sir James Goldsmith proposed an international company to be called "Forestco" to finance debt for nature deals up to $50 billion. Marilyn Post, "The Debt-for-Nature Swap: A Long-Term Investment for the Economic Stability of Less Developed Countries," *International Lawyer* 24 (1990): 1097.

25. See Lawrence K. Altman, "Fearful of Outbreaks, Doctors Pay New Heed," *New York Times*, May 9, 1989, B6.

26. The most closely related predecessor that I can find is Elisabeth Borgese's proposal for an "Ocean Development Tax." This was to include "a one-percent tax . . . on fish caught, oil extracted, minerals produced, goods and persons shipped, water desalinated, recreation enjoyed, waste dumped, pipelines laid, and installations built, all the major commercial uses of the ocean. . . . This functional, not territorial, tax would be levied by Governments and paid over to the competent ocean institutions (e.g., FAO, UNEP, IOC, IMO, International Seabed Authority) for the purpose of building and improving ocean services (e.g., navigational aids, scientific infrastructure, environmental monitoring, search and rescue, disaster relief, etc.)." *The Fu-*

ture of the Oceans: A Report to the Club of Rome (Montreal: Harvest House, 1986), 63. I review the history of this and similar proposals in Christopher D. Stone, "Towards a Global Commons Trust Fund," in *Freedom for the Seas in the Twenty-First Century: Ocean Governance and Environmental Harmony*, ed. John Van Dyke, Durwood Zaelke, and Grant Hewison (Island Press, 1993).

27. See "Announcement by President Nixon on United States Oceans Policy, 23 May 1970," *International Legal Materials* 9 (1970): 808.

28. UNCLOS Art. 160(2)(f)(i); and see Art. 140.

29. See, generally, Stoessinger et al., *Financing the United Nations System*, chapter 11.

30. In terms of legal doctrine, we would be arguing that the global commons is not subject to division analogously to a cotenancy under Anglo-American law, but that the wealth of the commons is inalienably the heritage of all the peoples of the earth forever.

31. Although as Onora O'Neill reminds us in *Faces of Hunger* (London: G. Allen and Unwin, 1986), we should not be too quick to blame these events wholly on Nature, without acknowledging the failure of human planning as a contributory cause.

32. In fact, how much to budget for assessing and valuing the risks, perhaps particularly at early stages of reaction, is itself a separate, expensive, and important task.

33. President Bush, remarks to the Intergovernmental Panel on Climate Change in Washington, D.C., on February 5, 1990, reprinted in "Two World Leaders on Global Environmental Policy," *Environment* 32 (April 1990): 12–14, 32–33. For a critical review of the program, including the suggestion that it appears "rather light on international cooperation," see A. Barrie Pittock, "Report on Reports," reviewing *Our Changing Planet: The FY 1990 Research Plan, The U.S. Global Change Research Program* and *Our Changing Planet: The FY 1991 U.S. Global Change Research Program*, by the Federal Coordinating Council for Science, Engineering, and Technology, Committee on Earth Sciences, *Environment* 32 (November 1990): 26. Of course, there is the risk that levels of domestic research on some areas will be inefficient—that with an adequate level of global cooperation, our own efforts would be better spent.

34. The WRI Report on Financing speaks of internationally harmonized rates of taxes on CFCs. World Resources Institute, *Natural Endowments: Financing Resource Conservation for Development* (Washington, D.C.: World Resources Institute, 1989), 22–23.

35. "Parties to Montreal Protocol Agree to Phase Out CFCs, Help Developing Nations," *International Environment Reporter (BNA)* 13 (1990): 275.

36. Of course, each contribution to a fund, once made, is also "sunk." The dollar value of the types of commitments must be compared.

37. See the discussion of Case (1) ("Internal") responses in chapter 2.

38. See "U.S. to help preserve resources in Guatemala with $10.5 Million Grant from AID," *International Environment Reporter (BNA)* 14 (1991): 148.

39. See Allen L. Springer, "United States Environmental Policy and International Law: Stockholm Principle 21 Revisited," in *International Environmental Diplomacy*, ed. John E. Carroll (Cambridge, U.K.: Cambridge University Press, 1988), 53, citing to "U.N. Holds Global Meeting," *Science News* 121 (1982): 358.

40. See chapter 3.

41. See chapter 4.

42. Based on 1987 data; the seven nations' contributions vary depending on which gases are counted. For example, the figure for CO_2 emissions only (cement, solids, liquids, gas, and flaring) is 58 percent. If the nations' methane emissions are included, the percentage is 52.56 percent. If East and West Germany's emissions for 1987 are now combined, it becomes the sixth largest polluter. See World Resources Institute et al., *World Resources 1990–91*, 346–47, table 24.1.

43. See Niala Maharaj, "Environment: Third World Suspects 'Green Fund'," *Inter-Press Service*, August 30, 1991.

44. Ibid.

45. Ibid.

46. See "Full Text of Kuala Lumpur Environment Declaration," *Japan Economic Newswire*, April 28, 1992. The delegates also "emphasize[d] that forest ecosystems and resources are part of the national patrimony to be managed, conserved and developed by each country in accordance with its national plans and priorities in the exercise of its sovereign rights."

47. When the seabed minerals regime was being actively debated in the UNCLOS negotiations, there was genuine concern that the administrative expenses of the proposed multitiered seabed apparatus would devour all the potential revenues, mooting the question of how to distribute any surplus among needy nations. See P. S. Rao, *The Public Order of Ocean Resources* (Cambridge, Mass.: MIT Press, 1975), 107–8.

48. Richard E. Benedick, *Ozone Diplomacy* (Cambridge, Mass.: Harvard University Press, 1991), 186.

49. See "Group of Legislators to Set Up Billion-Dollar Environmental Fund," *International Environment Reporter (BNA)* 15 (1992): 448.

50. There may be a sobering portent of this sort of thing in the ozone treaty. One provision invites nations (presumably LDCs) to notify the Secretariat if they do not have enough CFCs, see Article V, Section 4, *International Environment Reporter Reference File (BNA)* 21: 3156, with the perturbing implication that the treaty could wind up contributing to, rather than reducing, some nations' ozone-destroying activities.

51. The U.S. firms were increasingly hopeful, too, of being able to lead the way of getting acceptable substitutes to market.

52. Chapter 1.

53. See Lawrence K. Altman, "Fearful of Outbreaks, Doctors Pay New Heed," *New York Times*, May 9, 1989, B6.

54. Douglas Hunter Ogden, "The Montreal Protocol: Confronting the Threat to Earth's Ozone Layer," *University of Washington Law Review* 63 (1988): 1009.

55. M. Taylor and H. Ward, "Chickens, Whales and Lumpy Public Goods: Alternative Models of Public Goods Provision," *Political Studies* 30 (1982): 350–70.

56. On a related note, the United States is widely believed to be hampered in sponsoring fusion research, relative to its European competitors, because of the relatively short periods of budget review in the U.S.

57. One might suppose there is less room for controversy where the "matching" is done by reference to levels of domestic internal spending—-in other words, when nations can compare one another's commitments simply by comparing each of their budgets. But there will always be lingering uncertainty about how much each nation would have been spending on internal environmental programs, even in the absence of an agreement.

Chapter X
The Spiritual and Moral Dimensions

1. Lynn White, "The Historical Roots of Our Ecologic Crisis," *Science* 155: 1205.

2. Toynbee, "The Religious Background of the Environmental Crisis," *International Journal of Environmental Studies* 3 (1972): 143.

3. Yi-Fu Tuan, "Discrepancies between Environmental Attitude and Behavior: Examples from Europe and China," *Canadian Geographer* 12 (1968): 184.

4. Al Gore, *Earth in the Balance* (Boston: Houghton-Mifflin, 1992), 244.

5. See Malcolm W. Brown, "New Findings Reveal Ancient Abuses of Land," *New York Times*, January 13, 1987, A13: "Archeologists say idea of noble savage protecting the environment is a myth."

6. Tuan, "Discrepancies," 184.

7. René Dubos, *A God Within* (New York: Charles Scribner's Sons, 1972), 161.

8. Toynbee, "Religious Background," 144.

9. Even here, however, while one can find in Eastern religions more of a biocentric vision, knowledgeable Easterners caution about "the persistent invocation of Eastern philosophies" among Western environmentalists who do not appreciate the complex and fuller picture. See Ramachandra Guha, "Radical American Environmentalism and Wilderness Preservation: A Third World Critique," *Environmental Ethics* 11 (1989): 71–83.

10. I admit that the statement in text blithely ignores a considerable literature that would criticize me for exaggerating the ease with which market societies can control their corporations simply by escalating the price of wrongdoing. For a more skeptical view, see Christopher D. Stone, *Where the Law Ends: The Social Control of Corporate Behavior* (New York: Harper and Row, 1975); idem, "The Place of Enterprise Liability in the Control of Corporate Conduct," *Yale Law Journal* 90 (1980): 1–77.

11. I. Frolov, *Global Problems and the Future of Mankind* (Moscow: Progress Publishers, 1982), 116.

12. I examine the function of law in shaping culture in Stone, "The Force of Law in Shaping Cultural Norms Relating to War and the Environment," in *Cultural Norms in Relation to War and the Environment*, ed. A. H. Westing (New York: Oxford University Press, 1988), 64–82, and "From a Language Perspective," *Yale Law Journal* 90 (1981): 1149–92.

13. See Marshall Cohen, "Moral Skepticism and International Relations," in *International Ethics*, ed. Charles R. Beitz et al. (Princeton: Princeton University Press, 1985), 3–50.

14. A coalition that is more realistically estimated to number well over a hundred nations.

15. NIEO Resolution sec. 4(e), Sixth Special Session of the U.N. General Assembly, Resolutions 3201 (S-VI) and 3203 (S-VI), May 1974.

16. Quoted in Peter Bauer and John O'Sullivan, "Ordering the World About: The New International Economic Order," *Policy Review* 1 (Summer 1977): 5–69. Compare Peter Singer, "Helping is not, as conventionally thought, a charitable act which it is praiseworthy to do, but not wrong to

omit; it is something that everyone ought to do." Singer, *Practical Ethics* (Cambridge, U.K.: Cambridge University Press, 1979), 169. Onora O'Neill supposes Singer's view "is the central substantive ethical claim for any consequentialist"; O'Neill, *Faces of Hunger* (London: Allen and Unwin, 1986), 58.

17. W. Scott Burke and Frank S. Brokaw point out that a claimed "right" to the seabed wealth preempts conditioning distribution on ceasing human rights violations, etc. (for which reason Chile, Argentina, and Guatemala have suffered reductions in military aid). Burke and Brokaw, "Ideology and the Law of the Sea," in *Law of the Sea*, 45.

18. The concept was first brought before the United Nations in an address by Arvid Pardo, the Maltese ambassador, in 1967. It was endorsed in Resolution 2749 (xxv), The Declaration of Principles, 25 U.N. GAOR, Supp. (no. 28) at 24, U.N. Doc. A/8028 (1970), passed by a vote of 108–0 with 14 abstentions. For the subsequent history, including the U.S.'s disassociation from the moratorium resolution, see Steven Burton, "Freedom of the Seas: International Law Applicable to Deep Seabed Mining Claims," *Stanford Law Review* 29 (1977): 1147 and note 47. It is of interest that in 1969, Jack Stevenson for the U.S. proposed distribution of resources to the poorer nations, but, as is typical in diplomatic negotiations, felt unconstrained to adorn the offer with any philosophical formulation.

19. Iranian delegate, quoted in *Business Week*, "The Squabble over 'Parking' in Space," *Business Week*, November 8, 1982, 43–44. Subsequently, in a 1985 conference, a compromise was reached whereby each of the 160 nations in the organization was allotted a satellite position and frequencies. Details are still unsettled, and were slated to be taken up in 1988. See "Agreement on Satellites," *New York Times*, September 17, 1985, D22.

20. United Nations, General Assembly A/37PV/10, 29 September 1982, 17–20. In like vein, the Moon Treaty has designated the moon and its resources "the common heritage of mankind"—with mankind being understood here, as elsewhere in like treaties, to require, rather curiously, the principle of one nation, one vote, without any per capita weighting. See Allen Duane Webber, "Extraterritorial Law on the Final Frontier," *Georgetown Law Journal* 71 (1983) 1441 and note 104.

21. Whether the colonized were on balance disadvantaged by colonialization in a continuing, monetizable way is an open question. "Nor is it true that the per capita product of the less-developed regions declined during the period 1800–1950 because of 'colonial denudation' or from any other cause; on the contrary, it increased in some of the countries concerned, and

remained roughly constant in others [citing to S. Kuznets, "Population, Capital and Growth"]. It is true that the developed countries have been the primary beneficiaries of industrial progress, but this is hardly surprising, since they were its originators and the driving force behind it." Bernard Gilland, *The Next 70 Years: Population, Food and Resources* (Tunbridge Wells, U.K.: Abacus Press, 1979), 3.

22. Kenneth B. Noble, "Nations in Slave Trade Urged to Repay Africa," *New York Times*, August 10, 1992, A2.

23. "India: India Wants $2 Billion from Others to Sign Ozone Depletion Montreal Protocol," *International Environment Reporter (BNA)* 12 (1989): 389.

24. Trans. Ted Humphrey (Indianapolis: Hackett, 1983), 135 (p. 381 in original).

25. Some Kant scholars believe that the test places less emphasis on the actor being dissuaded by the sense of shame that application of the maxim would lead him introspectively to discover, than on the more practical consideration that if others learned what the actor was up to—if the practice were to be made public and was immoral—the actor risked being thwarted or punished. Kant was alive to the fact that people could quell feelings of shame if their interests were sufficiently engaged.

26. John Locke, *Second Treatise of Government*, sec. 27.

27. John Rawls, *A Theory of Justice* (Cambridge, Mass.: Harvard University Press), 377–79.

28. Rawls, *A Theory of Justice*, 7–8, 457.

29. Ibid., 378; Charles R. Beitz, "Justice and International Relations," in *International Ethics*, ed. Charles R. Beitz et al. (Princeton: Princeton University Press, 1985), 286.

30. Rawls, *A Theory of Justice*, 378–79.

31. Beitz, "Justice and International Relations," 287–88, suggests that Rawls felt that global commons issues would have to be left to treaty, on the view that the original international contractors, in legislating the basic framework, would focus on those issues designed to secure mutually desirable "internal" conditions.

32. Brian M. Barry, *The Liberal Theory of Justice* (Oxford: Clarendon Press, 1973), 133.

33. Beitz, "Justice and International Relations," 288. The unworkability of this distinction is discussed below.

34. Ibid., 138, 140. G. A. Cohen makes essentially the same point in "Self-Ownership, World Ownership, and Equality," in *Justice and Equality*

Here and Now, ed. F. S. Lucash (Ithaca, N.Y.: Cornell University Press, 1986), 108–35. I am not certain; while talent is—by definition—integral to the self, a natural resource may be an integral part of a nation, in the development of which a nation might take a special kind of pride. Imagine a leveling-out of central African water that resulted in taking the Nile from Egypt, even one that "compensated" Egypt with a market equivalent value in dollars.

35. Beitz, "Justice and International Relations," 305.

36. Ibid., 138.

37. Amartya K. Sen, "Ethical Issues in Income Distribution," in *Resources, Values, and Development* (Cambridge, Mass.: Harvard University Press, 1984), 280–82 and passim.

38. Robert L. Simon makes this point well in "Troubled Waters: Global Justice and Ocean Resources," in *Earthbound*, ed. Tom Regan (New York: Random House, 1984).

39. Let us grant that if we hold to a truly original position in which the covenantor has no idea in what generation he or she will be born, the chances of living through a (or the) generation in which wealth was reshuffled would be slight.

40. See Robert Nozick, *Anarchy, State and Utopia* (Oxford: Basil Blackwell, 1975). The suggestion that Rawls's original contractors would emerge with a system that embodied some of Nozick's more libertarian conception of rights is anticipated by John Stick in "Turning Rawls into Nozick and Back Again," *Northwestern University Law Review* 81 (1987): 363–416.

41. Readers seeking a superior technical treatment of the shadow price analysis which underlies the discussion in text are referred to Jean Dreze and Nicholas Stern, "Policy Reform, Shadow Prices, and Market Prices," *Journal of Public Economics* 42 (1990): 1–45.

42. Cass R. Sunstein, "Constitution and Democracies: An Epilogue," in Jon Elster and Rune Slagstad, eds., *Constitutionalism and Democracy* (New York: Cambridge University Press, 1988), 327–53.

43. See Mark Sagoff's discussion in *Economy of the Earth*, 14.

44. The Endangered Species Act claims to do this in some cases; but the political world is too pragmatic not to have inserted some *homo sapiens* favoring loopholes.

45. Gottfried Wilhelm Leibniz, *Theodicy*, trans. E. M. Huggard (New Haven: Yale University Press, 1952), sec. 118.

✤ Index to Topics and Institutions ✤

❖ Index of Authors ❖

✛ *Index of Conventions* ✛

✧ Index of Legal Cases ✧